Written and Illustrated

DESKTOP SCIENCE

by

B.K. Hixson

 Salt Lake City, UT 84115 • (801) 466-1172

Dedication

H. Dee Strange

Unfortunately, most of you who are currently holding this book in your paws have no idea who in God's green creation this woman is. To give you a general idea, take a rather small body, let's say one that is 4' 8" tall and all of 85 pounds if she were soaking wet in a very large sweatshirt. Age to near retirement and you would think that you'd have a sweet, demure, grandmotherly type of a lady who is great at baking cookies and never goes to an "R" rated movie. Wrong. Well, except for the "R" rated movie part.

Dee makes a hobby of traveling the world, visiting with stubborn border guards in Africa, eating walrus fat with the Eskimos, and riding in the back of a two ton truck for three days across the desert so she can say that she's been to Timbuktu. In fact, as this book is being prepared, Dee is in Russia riding the Trans-Siberian railroad and, because no one speaks English, she has been getting by speaking German. She was the "Teacher of the Year" at least one year in Hermosa Beach School District, and is always visited by her students (known as Strangemites) that have graduated. Mix in China, every National Science Teachers' Convention since World War II, several good causes ranging from working toward preventing the extinction of gorillas (before it was fashionable) to recycling aluminum cans, four kids, and tendencies towards being a devout Catholic, and you begin to get a hint of my good friend.

In addition to all of that, she is the woman who encouraged me to come back to education and hang out with kids rather than divert my attentions elsewhere. The joke in the whole thing is that if you were to go into her classroom, you would not find a desktop; there is too much junk. As you

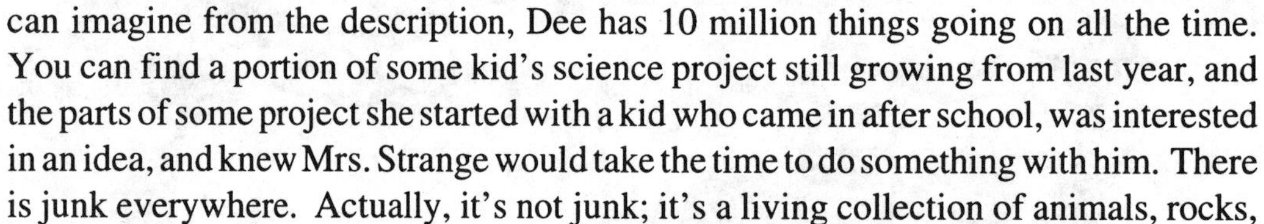

can imagine from the description, Dee has 10 million things going on all the time. You can find a portion of some kid's science project still growing from last year, and the parts of some project she started with a kid who came in after school, was interested in an idea, and knew Mrs. Strange would take the time to do something with him. There is junk everywhere. Actually, it's not junk; it's a living collection of animals, rocks,

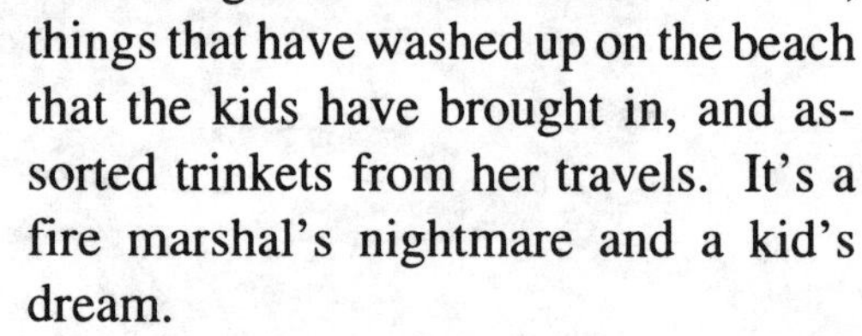

things that have washed up on the beach that the kids have brought in, and assorted trinkets from her travels. It's a fire marshal's nightmare and a kid's dream.

So, this is for you, Dee. Thanks for feeding me Big Macs on our trips to Joshua Tree. For sharing your enthusiasm for education and kids. I know that when I travel to Hermosa Beach to visit, there's a good chance that I'll see all 20 of these experiments going simultaneously, the way it should be. In closing, as you always say,

Onward!

Introduction

Welcome to Desktop Science. So much for the formalities. It is probably no great mystery to you that kids love to have their own individual experiments. They take an extra little bit of pride in the work and there is an increased interest in the projects. In the process of finding and developing experiments, we have come up with a collection with which we have had success in encouraging students to make daily observations, record data and mull over possible explanations for what they are seeing. The scientific method.

You'll find six sections for each of the lab activities. The first is the **Content Area**, which will get you going in the right direction when you are planning your year. The next is the **Learning Objective,** which zeros in on the specifics. The **Materials** that you will need are listed and that is followed with a step by step **Procedure**. As often as possible, without insulting your intelligence, I have made the directions painfully clear. My favorite section to write is called **Whyzat?** because too often we present ideas to students and they, quite naturally, ask why, I know that it's nice to have somewhat of a clue. You'll find the metaphors and explanations less conventional and hopefully more useful. The last section, called **Extensions,** consists of ideas for further exploration.

It never ceases to amaze me how many teachers will present an idea to their students without having ever tried the experiment on their own first. Usually something goes wrong due to this lack of preparation and I get a nasty letter. Don't do that; make like a kid and play. One of the best things that you could do for your students would be to perpetuate their sense of wonder and willingness to experiment. It is a well documented fact that by the time kids get to junior high they are so busy either trying to find "the right answer" or get the teacher's attention by setting the class gerbil on fire that they lose this very natural tendency. Encourage them to question, to invent, to explore, to wonder, and to understand that failure is a common and necessary element of life. After all, one of the few things that you can call your own are your mistakes. You will find that if you open this door for them, they will run through it at full speed. Kids love to play and that is what all those "brains" are doing in science labs all over the world....playing on their desktops.

Best to you in your teaching,

B. K. Hixson

Reproduction Rights

TABLE OF CONTENTS

The table of contents lists the lab activities in the order that they appear in the book. The lab activities are also grouped according to discipline. The first set is plant related labs, the second set, geology and so on. After the name of each activity there is a brief synopsis of the activity for quick reference. Happy experimenting!

Table of Contents

Materials List

If you are an ambitious soul and wish to assemble all of the materials that you will need for the entire book, this is your dream page. The materials are listed in alphabetical order. The number that immediately follows the material is the lab activity that calls for that material. This will make it convenient for you to assemble your own Desktop Science Kit. If you have difficulty finding some of the science oriented materials, give us a call and we can set you up.

2 liter pop bottles, 7
Acetone, 11
Adhesives (various), 16
Ammonia, 12
Apples, 20
Aquarium gravel, 6
Baby food jars, 6,13,20
Balloons, 9
Black felt pens, 11
Bowl, 11, 17
Bran, 20
Bread (white), 8
Cardboard, 5, 18
Cards (2 x 8, white), 17
Carnations, 3
Celery stalks, 3
Charcoal briquettes, 12
Chromatography paper, 11
Contac capsules, 19
Containers, 2,3,9,11,14,15,17,19
Copper sulfate crystals, 15
Disinfectants (various), 10
Egg cartons, 15
Eggshells, 15
Eyedroppers, 10
Food coloring, 3,12,13,14,17
Glass tubing, 17
Glue, 16
Grease pencil, 17
Hot plate, 10,13,15,17
Ice, 17
Laundry bluing, 12
Lima beans, 1, 6
Liquid plant food, 6
Masking tape, (almost every experiment)
Mealworms, 20
Milk cartons, 16
Paper plates, 4
Paper towels, 1
Paring knife, 2,10
Pencils, (all experiments)
Pie tins, 12,14
Pins (straight), 18
Plant specimens, 11
Plants, 2,4,5
Plastic bags (zip lock) 1, 8, 10
Potatoes, 2,10,20
Potting soil, 7
Refrigerator, 1
Salt, 12,14
Sand, 16
Sauce pan, 13,17
Scissors, 2,3,5,7
Seeds (assorted), 7
Shoeboxes, 5
Spoon, 11, 15
Stoppers (one hole), 17
String (cotton), 13
Sugar, 9,13
Test tube holders, 17
Test tubes, 9,17
Tongs, 10
Vegetable steamer, 10
Yeast, 9

LAB SAFETY

In every lab class there is always the danger that you may expose yourself to injury. The chemicals and equipment that you use and the way that you use them are very important, not only for your safety but for the safety of those working around you. Please observe the following rules at all times. Failure to do so increases your risk of accident.

1. Goggles.

Goggles should always be worn when chemicals are being heated or mixed. This will protect your eyes from chemicals that spatter or explode. Running water should be available. If you happen to get some chemical in your eye, flush thoroughly with water for 15 minutes. If irritation develops, contact a physician. Take this information with you.

2. Smelling Chemicals.

If you need to smell a chemical to identify it, hold it 6 inches away from your nose and wave your hand over the opening of the container toward your nose. This will "waft" some of the fumes toward your nose without exposing you to a large dose of anything noxious.

3. Chemical Contact With Skin.

Your kit contains protective gloves to wear whenever you are handling chemicals. If you do happen to spill a chemical on your skin, flush the area with water for 15 minutes. If irritation develops, contact a physician. Take the instructions for this kit with you.

4. Clean up all Messes Immediately.

This is no time to be a pig. Your lab area should be spotless when you start experimenting and spotless when you leave. If not, clean it.

5. Proper Disposal of Poisons.

If a poisonous substance is used or formed during the experiments in this lab, the book will tell you. These must be handled according to the directions in the lab guide.

Lab safety

6. No Eating During the Lab.

When you eat, you run the risk of internalizing poison. This is never done unless the lab calls for it. Make sure your hands and lab area are clean.

7. Horse play out.

Horseplay can lead to chemical spills, accidental fires, broken containers, and damaged equipment. Never throw anything to another person; be careful where you put your hands and arms; and no wrestling, punching, or shoving in the lab. Save that for when you get older and start dating.

8. Fire.

If there is a fire in the room notify the person in charge immediately. If they are not in the room and the fire is manageable, smother the fire with a blanket or use an extinguisher in an emergency and send someone to find an adult. REMEMBER: Stop, Drop, and Roll!

9. Better Safe Than Sorry.

If you have questions, or if you are not sure how to handle a particular chemical, procedure, or part of an experiment, ask for help from your instructor or an adult. If you do not feel comfortable doing something, then don't do it. If there is any concern upon chemical exposure, contact a physician.

PLANT EGGS

Content Area

Plants • Seed Germination
Plants • Parts of a Seed

Learning Objectives

1. The students will determine that seeds require sufficient water and the proper temperature to begin the germination process.

2. The students will dissect a seed in the process of germination and observe the parts of an embryonic plant. They will also identify and draw the seed coat, cotyledons, embryonic root and embryonic shoot.

3. The students may experiment with other techniques and ideas for germinating seeds to explore the diversity of options that exist for seeds to survive.

Materials You Will Need

4 Plastic bags (zip lock) per lab team
2 Paper towels per lab team
20 Lima beans per lab team
3 Extra lima beans per student
4 6" strips of masking tape / lab team
Refrigerator
Pencils or pens
Water

Procedure

1. Place half of the lima beans in a container. Fill the container with water so that it covers the seeds and let them soak overnight. The seeds will begin imbibing (soaking up) the water that they need to germinate. This step gives the seeds a head start. You'll want to add 3 extra seeds per student to the container. They are going to split the extra seeds in half and look at them (during the next several days) as they develop.

2. In addition to the 4 plastic bags and the 2 paper towels, give each team 10 seeds that have been soaking overnight and 10 seeds that have not.

PLANT EGGS

3. Ask the students to assemble each plastic bag following these directions:

Bag #1	5 "wet" seeds and a paper towel.
Bag #2	5 "wet" seeds and a paper towel.
Bag #3	5 "dry" seeds only.
Bag #4	5 "dry" seeds only.

4. Fold the bags over and seal each of them with a 6 inch strip of masking tape. Have the students write the following information on each bag. Eagle eye those little anklebiters to make sure that they put the proper information on the correct bag.

Bag #1	Name of students, wet, warm
Bag #2	Name of students, wet, cold
Bag #3	Name of students, dry, warm
Bag #4	Name of students, dry, cold

5. Have the students fill out the first day of their worksheet and then place bags 2 and 4 in the refrigerator. Let the students keep bags 1 and 3 on their desktops.

6. Now that the bags are set up, hand each of the students a lima bean that has been soaking in the water overnight. Ask them to split the lima bean in half using their fingers. Have them examine the lima bean and find the different parts of the seed that are identified on their worksheet. Discard the beans when you are done.

Whyzat?

This is the sequence of events that leads to the germination of the seed (with apologies to all of the hard core botanists and plant physiologists out there).

The plant prepares for its eminent survival next spring by producing seeds. In this seed it places a baby plant (called the **embryonic plant**) and food in the form of starch (the bulk of the seed) and covers it all with a seed coat to protect it from the environment. Once the seed is blown, dropped, or perhaps eaten by an animal and deposited in its feces, it lies dormant through the winter. In fact, many seeds require a period of severe cold before they will germinate.

Winter does its thing and then two things happen that get the seed's attention. One, the weather starts to warm up (which we all like) and two, it starts to rain

PLANT EGGS

(I'll let you have your own opinion on that one). This rain and the warmer temperatures cause the seed coat to soften and the seed imbibes (soaks up) the water that it needs to germinate and grow.

As the seed imbibes water, the embryonic plant wakes up and releases a hormone called **gibberellic acid**. This hormone then migrates to an area of cells called the **aleurone layer**, located just under the seed coat, which is where the next step kicks in. These cells have a concentration of protein molecules called enzymes that, when released by the gibberellic acid, begin the digestion of the starch stored in the seed for the baby plant. The embryonic plant can't use the starch as it is stored in the seed because the molecules are much too large. It would be the same thing as if you handed your 2 month old baby an apple. The poor kid would starve to death; it doesn't have the ability to cope with the food in this form, so you mash and squeeze the apple into pieces that the kid can digest properly. It's the same with the embryonic plant; the aleurone layer breaks the starch (which is a big, long molecule) into smaller bits we call sugar (much smaller molecules). The plant can use the sugars and oils (fats) to grow and develop. These parts of the seed are called the cotyledons, which means seed leaves. And to illustrate the point...

The embryonic plant releases gibberellic acid, which goes to the aleurone layer.

The aleurone layer cells are activated by the gibberellic acid.

The aleurone layer cells produce enzymes that digest the starch stored in the seed and the baby plant begins to grow.

As the seed begins to grow, the seed coat falls off and the embryonic plant emerges from the soil. Because the plant has not yet developed green leaves with chlorophyll to gather sunlight, it still relies on the stored energy that it started with, the cotyledon. Many people see these as the first set of leaves, but it is really the seed accompanying the shoot out of the soil. That's how it got its name—they are literally seed leaves. Clever people, those botanists. Anyway, as the plant grows, the food stored in the cotyledon is used up. Having served its purpose, the cotyledon shrivels up and falls off the plant. Kind of like the umbilical cord on a baby.

What the students will find in the course of their experiments is that only the seeds that are moist and warm will germinate. The seeds in the fridge that started out wet may begin germinating but will soon stop. Neither of the seeds that are dry will germinate.

PLANT EGGS

Extensions

There are several ways to modify and add variables to the experiment.

1. You can determine if the amount or kind of soil affects the rate of germination. You will find that there are all different kinds of soil from sand to loam to clay with different colors, acidities, and abilities to hold water. If you would like to add an English element to the unit, have the students write letters to different military bases around the country or the world asking for soil samples from the base. These guys are usually very good about writing back and sending samples. Determine rates of germination for several kinds of soil.

2. You can determine if the amount or color of light affects the rate of germination. Place plants in boxes that have had "windows" cut out of them. If you tape different colors of cellophane over these windows, you will find that plants do respond differently to different colors of the spectrum.

3. And my favorite, pollutants. For this one add one of several different kinds of pollutants to the soil with the water each day. You can try detergent, oil (in honor of the Valdez), ammonia, bleach, vinegar, or the goodie of your choice, in the quantity of your choice. Usually 1 to 3 drops applied directly to the seeds slows them up, and 5 to 10 drops completely wipes them out. The reason for the damage is that enzymes are very sensitive to acidity, or pH. By adding pollutants that are absorbed into the seed, you are tampering with the pH of the inside of the seed. It is kind of like taking your car key, mashing the heck out of it, and then wondering why you can't use it again.

4. One last idea before we move on. Take an old white, athletic sock (one that can run fast, just kidding) and put it over your shoe. Run around in a grassy field, remove the sock, put it in a plastic bag, add some water, and watch what happens. We call it a garden sock.

PLANT EGGS

*Name*________________________________ *Date*_______________

Observe the germinating seed each day. Draw a picture of what it looks like in the box and record the information in the spaces provided. Be as complete and accurate as possible. Pretend you are trying to describe these things to someone on planet Zorg who has never seen them before.

Days from start ________
Observations
Seed Coat________________________________
____
____
____

Root________________________________
____
____
____

Shoot________________________________
____
____

Days from start ________
Observations
Seed Coat________________________________
____
____
____

Root________________________________
____
____
____

Shoot________________________________
____
____

MORE PLANTS

Content Area

Plants • Asexual Reproduction
Plants • Propagation

Learning Objective

The students will observe the propagation of plants by growing leaf cuttings, germinating large existing seeds, or stimulating the growth of potato eyes.

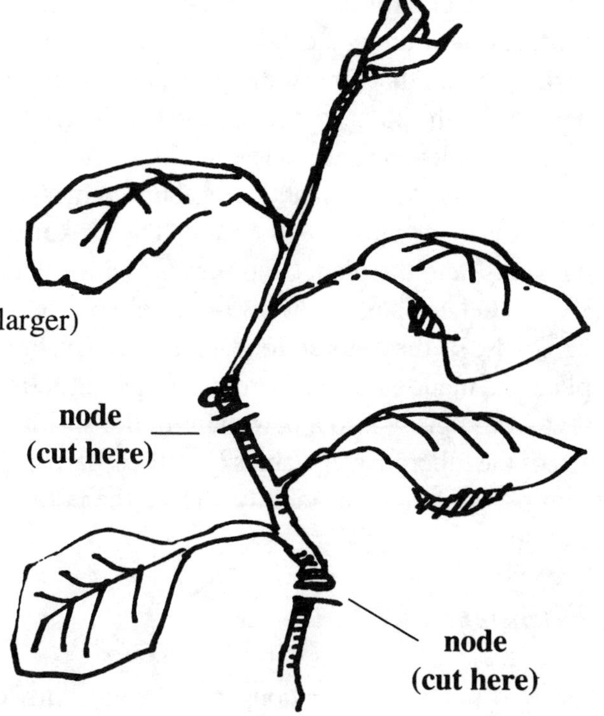

Materials You Will Need

1 Clear container/student (baby food jar or larger)
Paring knife (for the potato)
Scissors (for the herbaceous plants)
Masking tape
Pencil or pen

Any combination of the following:
Potato with eyes
One of these common house plants*
Coleus
Spider Plant
Wandering Jew
Philodendron
Violet

* If you have a green thumb, you'll know of several others.

Procedure

1. If you are using a herbaceous plant, take the scissors and cut the plant just below the node. Use the drawing at the upper right as an example. Place the plant in a jar that has water in it and label it with the name of the plant.

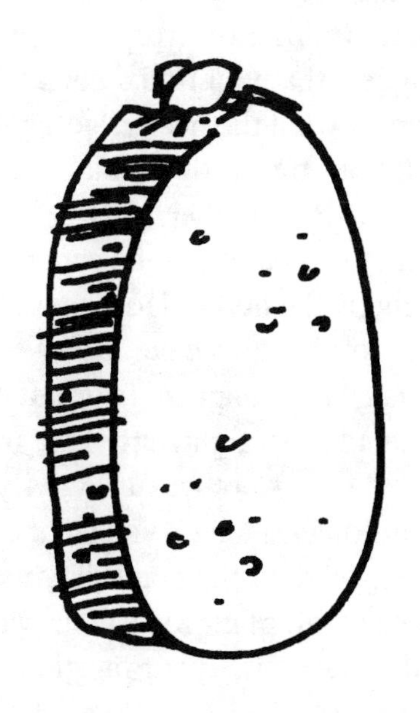

2. If you are using a potato, take the kitchen knife and cut on the sides of an eye, close to the eye, like you would cut a very fat potato chip for frying. Again, use the example at the right as a guide. Lean the potato slice in the jar, eye up, add water, and label the jar to complete the job.

MORE PLANTS

Whyzat?

A long time ago plants recognized that they were in a fairly precarious position as far as reproduction goes. I am certain that they didn't feel too secure in knowing that to ensure the preservation of the species, their only chance of survival was to whip up a bunch of seeds and let them zip out into the environment. They were hoping that, in addition to having several of them eaten, a few drug along for miles on the coat of some beast, and others washed down river, a scant few would find their way under a couple of inches of soil for the winter. These seeds would germinate the next spring and do that flower thing to start the process all over again. Seems a little chancy to me, too.

So, the plants got smart and invented other ways to reproduce. Knowing that they were going to get trampled, busted, and generally abused, they decided that when this happened, they would make the best of it. Many plants have a built in mechanism that triggers the growth of roots when the plant is broken in the middle of the stalk. The broken end is elected to be the new base of the plant and sends out roots, and the plant continues growing. These are our herbaceous plant specimens.

Other plants didn't want to rely entirely on seeds, so they stored food in the root of the plant and made sure that the root had special cells that would grow into a new plant in the spring. All of the tuberous vegetables (potatoes, carrots, turnips, etc.) have this ability. Thus, the potato example.

Extensions

1. The most obvious (and the most difficult to accomplish) would be to experiment with grafting two plants together. This works best with fruit bearing trees. If this is of strong enough interest to you, I would suggest that you find a book written by someone who knows what the heck is going on because I'm sure I'd kill your tree with my advice.

2. Another activity would be to experiment with the location of the cut. Does it always have to be right by the node? How about the middle of the plant? Do you have to have leaves for the plant to survive? What would happen if you cut the plant stem vertically instead of straight across the middle?

3. Another fun activity would be to see if the solution into which you put the leaf cutting influences the growth of the plant. Try using water, sugar water, water with plant food in it, water with food coloring, salt water. Let your imagination run wild.

MORE PLANTS

*Name*________________________________ *Date*_______________

Observe the plant daily. Record changes in the growth and development of the plant in the spaces provided.

Days from start ________
Observations
Root: ______________________________

Shoot: ______________________________

Days from start ________
Observations
Root: ______________________________

Shoot: ______________________________

WATER TUBES

Content Area

Plants • Transpiration
Plants • Capillary Action

Learning Objective

The students will observe and describe the movement of water through the vascular tissue of celery stalks and/or carnations.

Materials You Will Need

Food coloring
Water
Container that will support celery stalk per student or group
Celery stalks or carnations per student or group
Scissors or knife

Procedure

This activity is best if done early in the morning and observed throughout the day.

1. Have the students fill their containers two thirds full with water and add food coloring to the desired pigment intensity. Darker colors obviously work better, but I would let the students explore that idea if you have time.

2. When everyone is ready, cut the hearts of the celery stalks off with the kitchen knife and immediately place the stalks in water. In fact, if you cut the hearts off under running water, the vascular tubes in the celery (or carnation) that carry the water will not have any time to collapse and the students will get better results.

3. Have the students bring their containers up to the front of the room and choose a celery stalk to place in their colored water. Ask them to make the transfer between the two containers quickly and return to their desk to begin making observations.

WATER TUBES

Whyzat?

The students will notice during the course of the day that the colored water gradually works its way up the stem. Here's how it works. The water the plant needs to live and grow is in the ground, so the plant sends its roots down into the soil to retrieve it. Once the water is in the roots, it is transported to the stem of the plant where it is pulled into water canals called **vascular tubes**. The water is then distributed throughout the plant.

The big mystery here is what makes the water move through the plant. Is it pushed, pulled, or what? As it turns out, the whole process is started in the leaves through lots of little holes called **stomata**. One of the purposes of these holes is to allow for the escape of water in a process called **transpiration**. As water is lost through the leaves, replacement water is pulled from the stem, which then pulls water up from the roots. This pulling process is called **capillary action** and is the result of the attraction water molecules have for each other—kind of like little magnets.

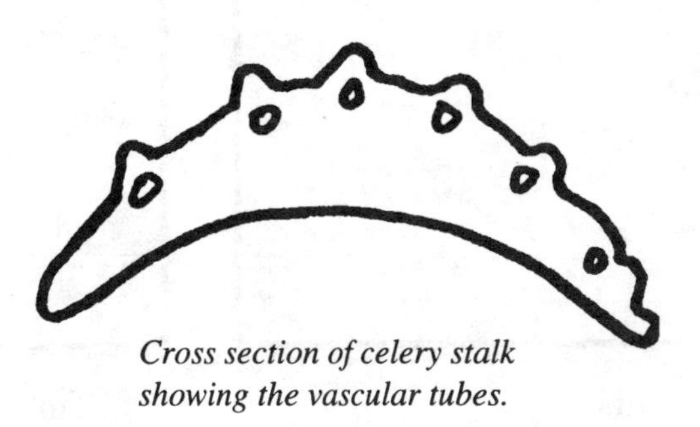

Cross section of celery stalk showing the vascular tubes.

The water in these stems also provides support for the plant, which is why plants wilt if the soil dries out. Another way to think about it is like the big fire hoses used to combat fires. When they are empty, they are easy to move around but when they are full, the hose is taut.

Extensions

Try these different combinations and compare the difference in the rate of transpiration:

1. Put one plant in the sun and the other in the shade.

2. Put cellophane around the bottom of one stalk of celery and none on the bottom of the other. If you want to be even sneakier, just rub a little petroleum jelly on the cut ends of the stalks.

3. Add salt to the water of one plant and keep the other in fresh water. If that does not interest you, try sugar, vinegar, cooking oil (good luck mixing that one), bleach, or any other thing your heart desires.

4. If you take a white carnation and put it in colored water for a while, you'll see that the food coloring is left in the petals of the flower. Holiday color for your flowers.

5. Transfer the flower from container to container and the colors of the petals will show colored bands. This is good for school colors, patriotism, or amusement on a really slow Friday night.

6. Split the base of the stalk of the carnation up the middle and place one half in one jar of colored water and the other half in another jar of water that is a different color.

Water tubeS

*Name*______________________________ *Date*_______________

Observe your celery stalk every hour. Record the time and draw a picture of how far the colored water has moved up the stem.

Time ____________

Time elapsed ____ hrs.

Time ____________

Time elapsed ____ hrs.

Time ____________

Time elapsed ____ hrs.

Time ____________

Time elapsed ____ hrs.

Time ____________

Time elapsed ____ hrs.

Time ____________

Time elapsed ____ hrs.

RIGHT SIDE UP

Content Area

Plants • Geotropism

Learning Objective

The students will observe how a plant that has been tipped on its side compensates for the pull of gravity.

Materials You Will Need

1 Small 6" plant in a pot (coleus or beans both work well) per student
(You'll need pots, lima beans, potting soil and water.)
1 Paper plate per student

Procedure

1. A couple of weeks before the lesson (2 to 3 depending on the size of the plants that you want to use), plant and grow lima beans for the students. Grow 1 plant for each student.

2. Give each student a plant and a paper plate. Have the student tip the plant on its side and draw a picture of what they see. This activity will take a couple of days for them to finish so have them set the plate on their desktop and take out that history book. After all, a watched plant doesn't geotropify.

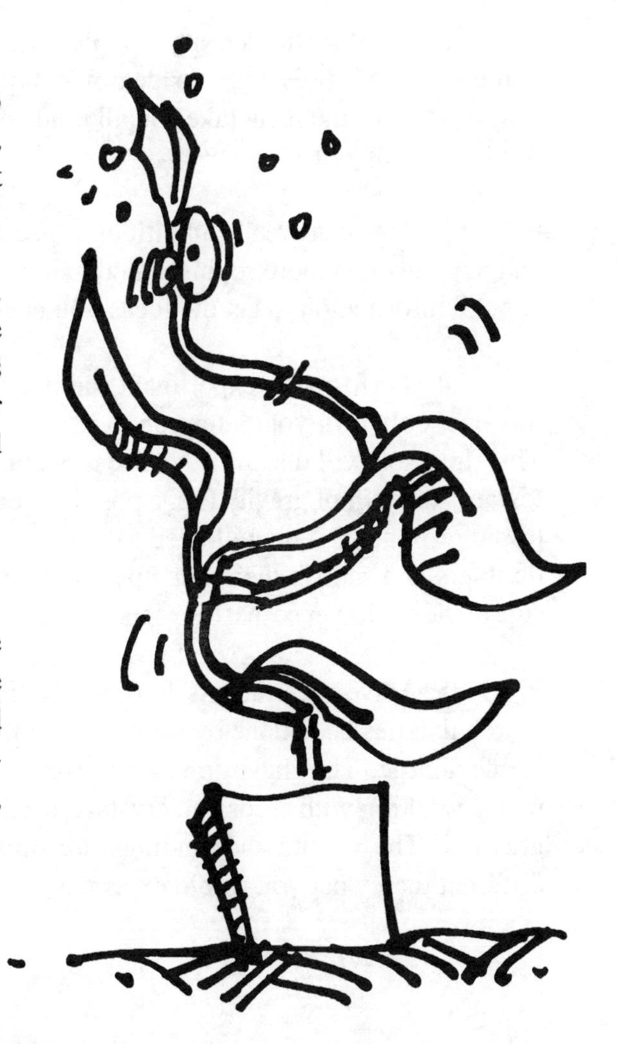

Whyzat?

Plant hormones are directly responsible for the amount and direction that a plant grows. They control the length of the stem, the bushiness of the leaves, all that good stuff. The hormones themselves are subject to temperature, light, and gravity. In this particular experiment, gravity is the focus of our interest.

RIGHT SIDE UP

A plant hormone, called an auxin, is responsible for the elongation of the stems in plants. When a plant is placed on its side, those hormones migrate laterally to the downward side of the stem. Because there is a concentration of the auxins on the underside of the stem, the cells there elongate, grow, and generally right themselves so that the stem turns away from the pull of gravity (i.e. a negative geotropic response). It amazes me, too. It is not clear if the auxins move because of light or gravity in this particular case. They are light sensitive, but there are other experiments that indicate that gravity is indeed responsible for the stem movement. You will probably wait until the kids get to high school or college before they dip into that kind of stuff.

Extensions

1. If you have an old phonograph or are careful to cover the one that you have, it's a lot of fun to put a fairly tall plant on the turntable and let it grow while the turntable spins. If you let it go for a couple of days, the plant will eventually lean to the center of the turntable. For best results, use the lowest speed; 16 is optimal.

2. Another fun activity is to play "mess with the plant." Every 3 or 4 days, change the position of the plant. Move it from upside down, lay it on its left side, then its right side, upside down again, and the stem of the plant takes on all kinds of weird contortions. One word of caution: don't let the S.P.C.P. catch you.

3. If you are really ambitious, you can get auxin paste from the science supply store and work on the plants in a more sophisticated manner. Any introductory college botany text would give you enough information to get the ideas you need to turn your imagination on.

4. Perform an experiment similar to this one, only turn your attention to the roots. This time you will discover that the roots turn towards the pull of gravity (i.e. a positive geotropic response). You will need to have clear tumblers and moosh the seeds up against the side of the container so that the roots are visible.

5. Have the students look up the research that has been done by astronauts in the Space Shuttle. They have done several experiments working with seeds and crystals in zero gravity. The results and findings are quite different than what you would expect.

Right side up

*Name*______________________________ *Date*_______________

In the spaces provided, draw a picture of your plant each day. Describe any changes that you have observed in the stem and leaf position.

Days from start ________
Observations
Stem ________________________________

Leaves ________________________________

Days from start ________
Observations
Stem ________________________________

Leaves ________________________________

WHERE'S THE LIGHT

Content Area

Plants • Phototropism

Learning Objective

The students will observe how a plant moves through a maze toward light, demonstrating the inherent phototropic response.

Materials You Will Need

1 Small plant in pot (lima beans work great)/student
1 Shoebox (preferably empty with lid)/student
Cardboard
Masking tape or glue
Scissors

Procedure

1. Germinate those lima beans 3 weeks in advance and make sure they are at least 4 inches tall. If you are inspired and lack for pots, use old discarded milk cartons. There are usually one or two in and around the cafeteria on a good day.

2. Have each student bring an empty shoebox from home. Their job is to design a maze for their plant to travel through. There is an example on the next page in the cartoon illustration. I would suggest having the students help you design several "floor plans" as a class, and then give them some cardboard, tape, and scissors and let them have at it individually, stressing the individually part.

3. Once the mazes are constructed, have the students cut a hole, roughly 2 inches in diameter, in the end of the shoebox. This will allow light into the box and give the plant something to grow toward.

4. With the box in a horizontal position, place the plant in the far end of the box, away from the opening. You may have to start the plant into the maze to put the lid on the box. This is not "cheating" as some of the students will contend. Put the top on the box and write the names of the group members on the lid.

WHERE'S THE LIGHT

5. Once the mazes are ready, have the students fill in the record sheets and begin observations. Tell the students that the project will extend over several weeks and not to get too anxious if the plants do not grow at the speed of light toward the opening in the box.

6. It seems that because we are concerned about light in this experiment the students forget about water. Be sure to have them check their plants every couple of days.

Whyzat?

As was discussed in the activity just prior to this one, plants are subject to the movement and influence of hormones. Our buddy **auxin** is responsible for the movement of the plant toward light in this experiment.

Auxins are a group of hormones, many of which are light sensitive, that stimulate the elongation and growth of plant cells. If a plant is in the shade of another plant, the auxins migrate to the darkest side of the plant and collect there. This concentration of auxins causes the cells on the dark side of the plant to grow and elongate, bending the plant toward the light. This is why the plant has the ability to "grow" around corners. It's just the shifting of the auxins back and forth in the stem.

Extensions

1. The tip of the plant is one of the sources of the auxins in the plant. Cut the tip of the plant off and see what happens when the plant is placed in the maze.

2. Design a vertical maze and see if that has any influence on the movement of the plant.

3. Design a container that allows you to see the roots of the plant. Carefully wash the soil from the roots, cut the tips of the roots off, and replant the plant in a see-through container. Record your observations.

4. Use squares of colored transparency film to form the barricades in the box. They may need to be reinforced with cardboard, but the main idea is to have the plant think its way around the colors.

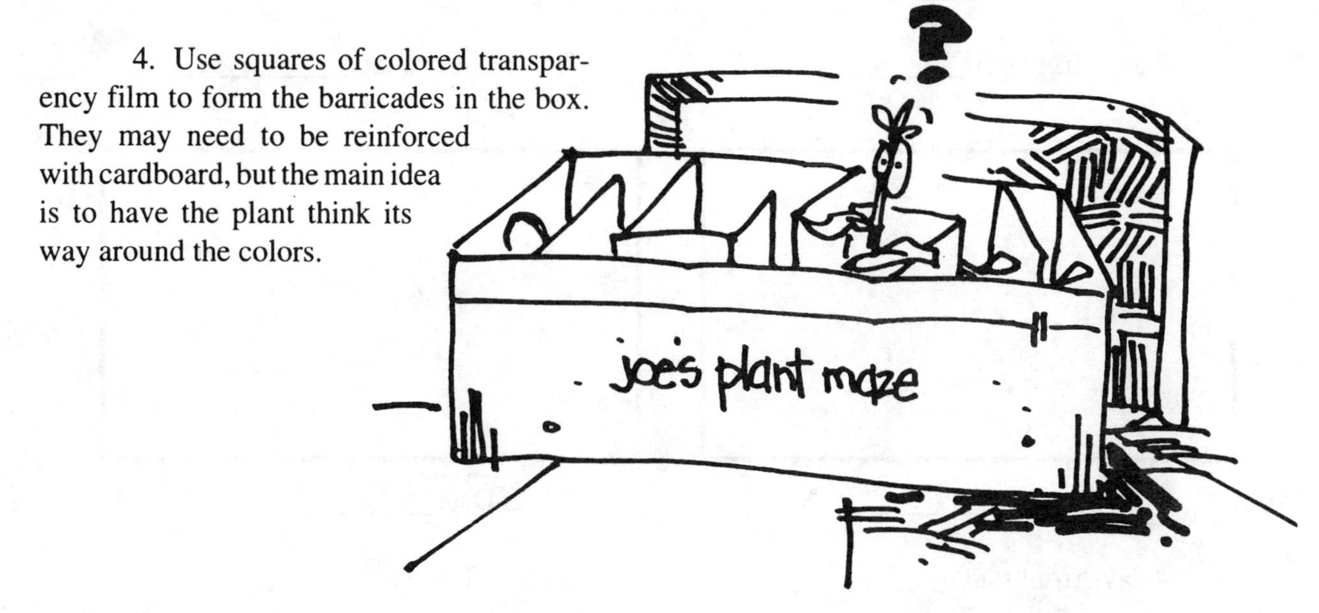

WHERE'S THE LIGHT

*Name*________________________________ *Date*______________

Draw a top view of your maze and record the growth of the plant.

Date ____________

days from start ____________

Date ____________

days from start ____________

Date ____________

days from start ____________

Date ____________

days from start ____________

Date ____________

days from start ____________

Date ____________

days from start ____________

JUST WATER

Content Area

Plants • Hydroponics

Learning Objective

The students will germinate, grow, and observe plants without the use of soil.

Materials You Will Need

3 Baby food jars per lab team
Aquarium gravel or vermiculite
6 Lima beans per lab team
Water
Liquid plant food
Masking tape
Pencil or pen

Procedure

1. Give each of the lab teams 3 baby food jars, tape, 6 lima beans, aquarium gravel or vermiculite, and a pencil, if they don't have one.

2. Have the students fill each jar two-thirds full of gravel or vermiculite and then plant 2 seeds smack in the middle, about a half an inch down. Be sure the seeds are covered.

3. Label each of the jars with tape and have the students prepare each jar according to the label they have just put on it. In adding the water, make sure that it comes up to the level of the gravel initially.

Jar #1 no water
Jar #2 water
Jar #3 water with liquid plant food (5 drops or per directions)

As the plant grows and the roots develop, you will need to add less water. You can actually drown a plant if you give it too much water, so have the students be careful.

JUST WATER

4. You may want to mix the plant food before class rather than have each of the lab teams mix it at their tables; it's up to you. That's why you get the big bucks.

Whyzat?

This branch of science is called **hydroponics**, or the science of growing plants without soil. The soil provides two things for the plant: a substrate or base for the plant to grow in and minerals for nourishment. It goes without saying at this point that the seeds in container number 1 will not germinate because they do not have the necessary water. The other two will germinate, but number 2 will eventually die (or, at best, turn a very pale green and be stunted in growth). Only the seeds in container number 3 will flourish over the long haul.

The vermiculite (or aquarium gravel) acts as the substrate and provides the base for the plants. As the seed germinates, the roots grow into the substrate and use it for support. At this point the plant doesn't really care if there is soil or not, as long as there is something to hang on to. There are lots of set ups that don't even rely on gravel as the substrate; the plants are supported by other means.

The nutrients that the plant needs in the early going are provided by the cotyledons (also called "seed leaves"). Once the energy in the seeds is used up, the plant had best have its root system in working order or perish. In this particular instance, the nutrients the plant needs will come from the plant food solution that you add to container number 3. The other two containers will not have this source of nutrient; the results should show this quite clearly.

Extensions

1. Challenge the students to figure out a way to support the plant in a solution of nutrients and water without the use of vermiculite or aquarium gravel.

2. It is possible to grow vegetables hydroponically to maturity. If there is an ambitious soul in your class that would like to take on this challenge, have them design, layout, and plant a hydroponic garden.

3. Challenge the students to create a mini-greenhouse that is oriented around hydroponic principles.

JUST WATER

*Name*____________________________ *Date*______________

In the spaces below, draw your plant. Describe the growth of the plant, the color of the leaves, and any changes for each of the 3 experiments. These are side view drawings.

Observations Date__________

MINI GREENHOUSE

Content Area

Plants • Greenhouses
Plants • Terrariums

Learning Objective

The students will construct a terrarium from a plastic, 2 liter bottle and observe the growth and changes that take place.

Materials You Will Need

Potting soil
2 liter pop bottle per lab group
Assorted seeds (grass, radish, bean, etc.)
Scissors
Masking Tape
Pencil or pen

Procedure

1. Cut the top of the bottle off and recycle it.

2. On the bottom of the bottle there is a black support stand. Wiggle it off under hot water and partially fill it with potting soil. Plop in an assortment of seeds and then cover them with about a half an inch of potting soil.

3. Moisten the soil and place the clear top of the container upside down on the black support stand. Put a piece of tape on the black bottom and record the names of the members of the lab team and the date the greenhouse was started.

4. Have the students water the greenhouses if they appear to be losing moisture and record the growth of the plants. They may want to add a snail or a bug to the terrarium just to give the local zoo a little competition.

5. As the students make their observations, be sure to encourage them to describe the colors of the leaves, the sizes and shapes of the plants, the changes any of the insects may make, and the amount of moisture that is being transpired into the container.

MINI GREENHOUSE

Whyzat?

The reason the greenhouse is able to survive without much attention is because it is self-contained. The water does not really evaporate from the area, it just collects on the inside of the container and then trickles back down into the soil. The plants transpire, but again the water is collected on the inside of the container and trickles back down into the soil to be recycled.

This is actually a very accurate model of the Earth as a whole, because we never let any water actually escape from the atmosphere. There may be dry spells, but the water never leaves the planet.

Extensions

1. Ask the students to study the different ecosystems and recreate as many of them as is possible. Deserts, rain forests (ever so politically correct), grasslands, forests, cities (not official ecosystems, but what the heck), tundras, oceans, rivers, streams, and estuaries come to mind as plausible possibilities.

2. Build two identical "ecosystems." Keep one covered and the other uncovered; record the results.

3. Have the students build an aquarium. Record or make observations as to the differences between a terrarium and an aquarium.

4. Choose a type of ecosystem as a class and turn the entire room into a replica of said ecosystem. The example that comes to mind is that my friend, Paul Quinby, who teaches in Southern California, and his students made a rain forest complete with the large trees, vines hanging down, and recorded animal noises. They even had a thing called a water stick (available at Natural Wonders and other fine retail outlets of ecological paraphernalia) and had rain showers at regular intervals.

5. Visit the local zoo. If it is any good at all, it will have the different habitats, biomes, and ecosystems on display with resident animals. In some cases scheduled rainstorms and enclosed spheres with extremely high humidity are also available for temporary discomfort of an educational variety.

MINI GREENHOUSE

*Name*________________________________ *Date*________________

Record the growth of the plants in your greenhouse in the spaces provided. Draw both a side view and a top view (remove the top of the terrarium if you need to).

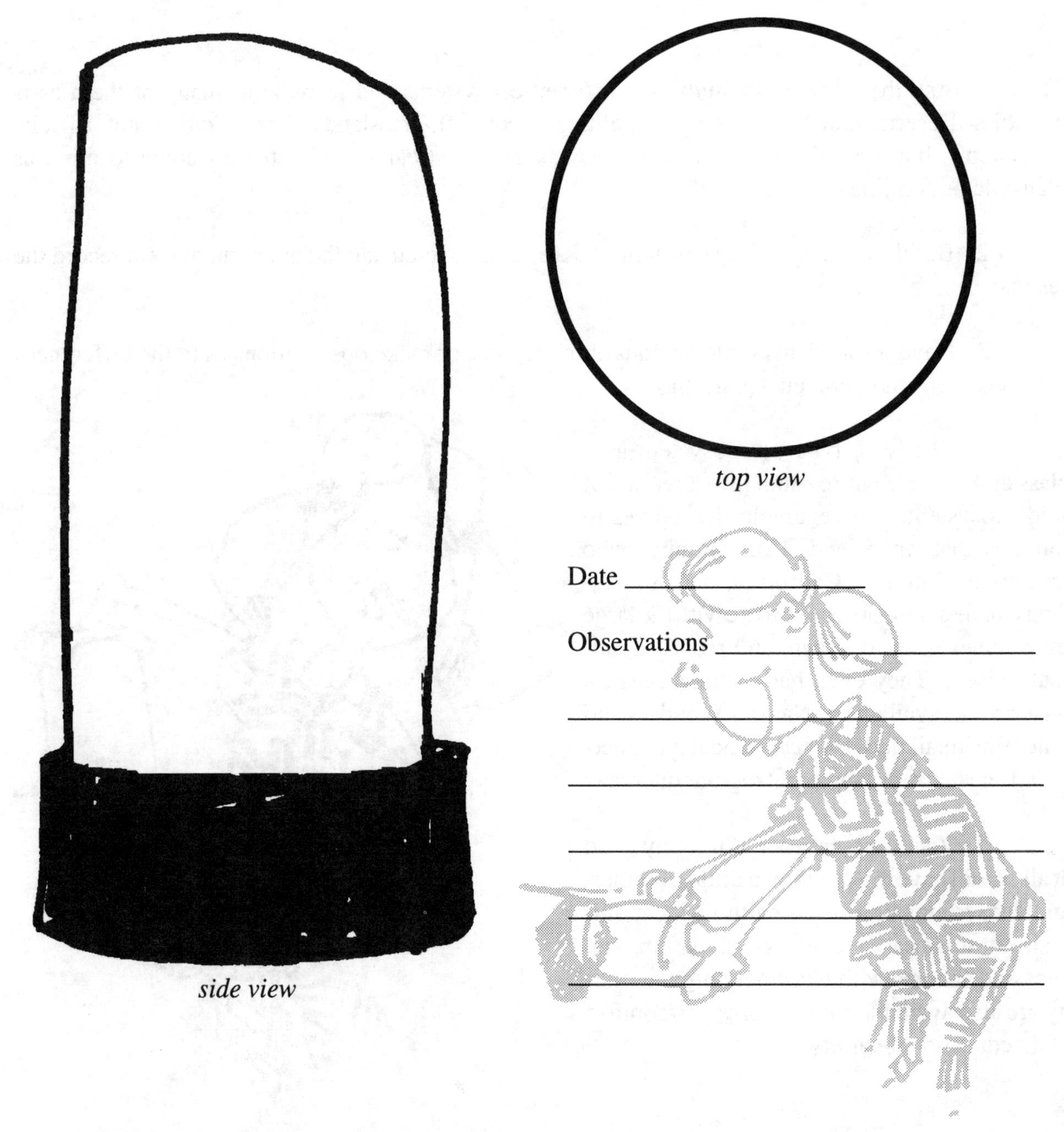

side view

top view

Date ________________

Observations ______________________

____

____

____

____

____

GOOEY BREAD

Content Area

Decomposers • Mold

Learning Objective

The students will grow and observe a mold culture on a piece of bread.

Materials You Will Need

4 Plastic bags (zip lock)/ team
4 4"x4" chunks of white bread (the colors show up better)
Masking tape
Pencil or pen

Procedure

1. Hand each of the lab teams their materials and instruct them to label each of the bags in the following manner:

Bag #1	Bread control
Bag #2	Bread with water
Bag #3	Bread wiped on the floor
Bag #4	Bread with saliva on it

2. Prepare each of the bags using the information above. Bag #1 is the control so they are to just put the bread in the bag and seal it. To the others, add the items that are listed on the labels.

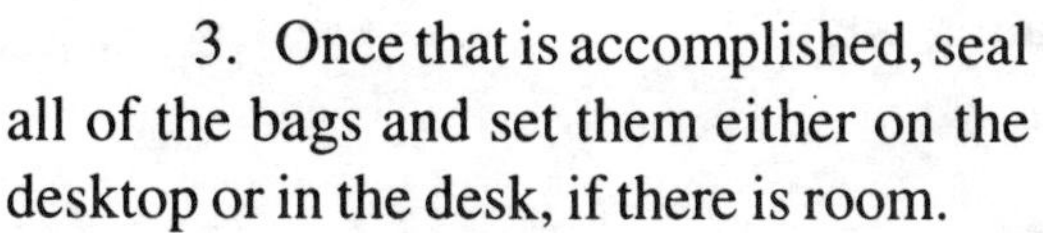

3. Once that is accomplished, seal all of the bags and set them either on the desktop or in the desk, if there is room.

GOOEY BREAD

Whyzat?

The mold that you are seeing is called **rhizopus** (rise • oh • puss); it's common, black, bread mold. The structure of rhizopus is quite amusing. It looks like a giant red playground ball on a spindly stalk anchored with a couple of roots. The bread mold matures and produces spores in the ball at the top. When the **spores** (mold seeds, for lack of a better description) mature, the walls become very thin and rupture, releasing the spores into the air so they can travel and land on a new chunk of territory and start all over again. This is what you see when you blow on the mold and see all of that "dust" flying up into the air. It's really spores flying around.

The additional tests, using the saliva, water, and dust off the floor may yield some other critters that are interesting, but the purpose of this lab is to grow rhizopus goodies to look at. It is best if the bags remain sealed at all times. The spores won't kill you, but they aren't great for you, either.

Extensions

1. Experiment with the effect of temperature on the growth of the mold colony. One of the great advantages of modern refrigeration is that it retards the growth of mold and bacteria. As you are well aware, they tend not to thrive very well in the cold... probably part Hawaiian or something. Anyway, if you place a piece of bread in the fridge, one on the table, and another in the furnace room (with the janitor's permission, of course), you'll find dramatic differences in the rates of the mold growth. The general law for chemists is that the rate of the reaction doubles for every 10 degree (Celsius) increase. My guess is that the biochemical reactions that promote the growth of these organisms are governed by a similar rule of thumb.

2. If a microscope is available, it is a lot of fun to look at the bread mold in a magnified state. The students will see the **sporangia** (the ball at the top) and be able to identify the **sporangiophores** (stalks) and **rhizoids** (roots) as well. If you have careful scientists they will be able to collect spores from the sporangia and look at those, also.

3. Experiment with other kinds of bread and different sources for mold (e.g. cheese, yogurt, and various kinds of fruit). Anything that is composed of organic matter makes a good substrate for mold and bacteria. My only caution to you would be to have the students wrap everything and keep it sealed unless you know for sure what it is that you're dealing with.

4. Commercial agar is available for purchase from a number of science supply houses. Agar is scientist's Jello (made from seaweeds, mmm, yummy) that has been especially prepared for growing microscopic goodies. It is sterile and can be used by the students to show the value of washing hands, brushing teeth, or just collecting goodies from the counter.

Experiment 8

Gooey Bread

*Name*____________________________ *Date*______________

Draw a picture of each of the 4 bread samples. Describe the colors, shapes, and textures (fuzzy, smooth, rough, stringy) of each sample.

1.

2.

3.

4.

BREAD BUBBLE MAKER

Content Area

Microorganisms • Yeast
Life Cycles • Carbon Dioxide

Learning Objective

The students will experiment with different combinations of water, sugar, and yeast to see how they influence the production of carbon dioxide by yeast organisms.

Materials You Will Need

Yeast
Sugar
Water, warm
4 Test tubes per lab team
4 Balloons per lab team
Container that will hold the test tubes
Masking tape
Pencil or pen

Procedure

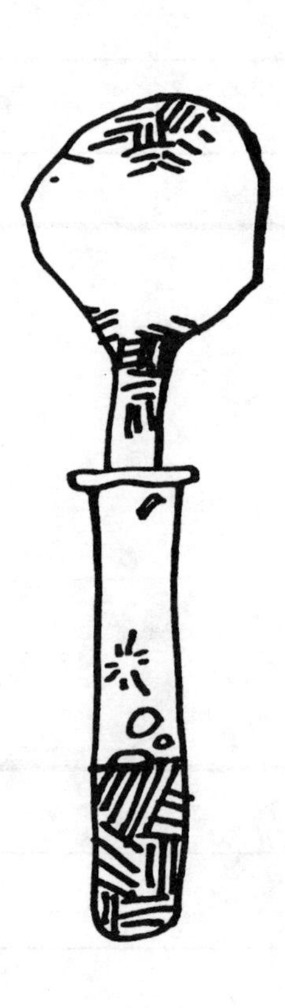

1. Label the test tubes 1 through 4 with pieces of masking tape.

2. Prepare each of the tubes using the following quantities:

Water • one half a tube (use warm water)
Sugar • one pinch
Yeast • one pinch

Tube #1	water, yeast, sugar
Tube #2	water, yeast
Tube #3	water, sugar
Tube #4	yeast, sugar

3. Place a balloon over the top of each test tube and store them upright in the container.

BREAD BUBBLE MAKER

Whyzat?

Yeast is a living organism, and as such it eats, digests, and excretes waste products just like we do. Another similarity between yeasties and us is that yeast eat sugar and produce carbon dioxide as a waste product. Coincidentally, we breathe oxygen and produce carbon dioxide as a waste product. I can imagine that the hard-line creationists are getting the old hemp necktie out about this time thinking that I'm going to tell them that they, quite obviously, evolved from yeast. I won't. I will leave it to you to determine your own origins.

Back to the subject at hand. The yeast are eating the sugar that you put in the test tube, the bread, and anywhere else. As they eat, they produce carbon dioxide. The carbon dioxide collects in the bread dough and this is what is responsible for the holes that we see in our bread. If one of the components is absent, the yeast will not be able to live and there will be no carbon dioxide produced. Therefore, we get unleavened bread for Passover.

Extensions

1. Make bread with and without yeast.

2. Vary the amount of time you allow the yeast to rise and compare the two loaves of bread (e.g. hole size, texture, etc.). Have fun with this one.

Evolution clarified during an argument.

Bread bubble maker

*Name*______________________________ *Date*_______________

Observe each of the 4 test tubes every hour for 6 hours. Describe the colors and changes, and draw the test tube and balloon for each experiment.

Observations

BACTERIA FARM

Content Area

Decomposers • Mold and Bacteria
Decomposers • Disinfectants

Learning Objective

The students will use mold grown on a potato template and bacteria collected from their mouths. The effect of various disinfectants will be tested on these growths.

Materials You Will Need

Vegetable steamer
Hot plate
Tongs or a pair of forks
Water
5 Plastic bags (zip lock) per lab team
(or petri dishes, if you have them)
Several large potatoes
Paring knife
A variety of liquid disinfectants, cleansers, and mouth washes
Eyedroppers
Pencils or pens
Masking tape

Procedure

1. Slice the potatoes into sections that are one-half inch thick. You can slice them thinner if you would like, but once they are steamed they tend to fall apart easily.

2. Get the steamer up to speed before you put the potato slices in. Cook them for 10 minutes, 15 if you are trying to kill time.

3. As the potatoes are pulled out of the steamer, have the students ready with clean plastic bags or petri dishes and place them in a closed environment immediately. Avoid exposure to the air and germs as much as is possible.

Bacteria farm

4. Label the plastic bags 1 through 5 using the masking tape and pencil. Keep bag number 1 as a control (i.e. no disinfectant added). Identify the disinfectant to be used on each of the remaining potatoes.

5. Once the plastic bags are prepared, have the students spit, lick, or drool on all five of the potato slices. Given the travels of most students' mouths, this should contaminate the potatoes nicely.

Bacilli

6. Once the contamination is complete, have the students add 10 drops of the disinfectant marked on the plastic bag to the potato surface. Make sure to tell the students that the drops should be concentrated on one half of each of the slices.

7. Set the potato slices aside and let the students check them once a day and record the growth of the different goodies. Don't open!

Whyzat?

The fact that you are able to grow bacteria so easily is no coincidence. They are the most abundant living organisms in the world, living in extremes from the Antarctic to the near boiling waters just out of the mouth of hot springs and geysers. The reason for their success is that the population of a colony can double every 20 to 30 minutes, if the conditions are right. Just look in your fridge.

Cocci

Bacteria come in three different shapes: the bacilli (buh • sill • eye) or rod shaped bacteria, the cocci (cocks • eye) or round bacteria, and the spirilla (spy • rill • uh) or the spiral shaped bacteria. There are somewhat accurate drawings of each group to the right. No, bacteria do not have eyes, mouths, or teeth.

The potato acts as a friendly environment for the bacteria. There is plenty of food in the form of starches and sugars, a nice flat place to land, and there are no competitors (as of yet) when the initial cells land on the surface because the potato was pulled out of a steam bath that killed everything. When the students licked or spit on the surface of the potato they inoculated it with the bacteria in their mouths, which set the whole process of contamination in motion.

Spirilla

BACTERIA FARM

The disinfectants are responsible for the removal or prevention of bacterial growth. As you will witness, there is a wide variety of effectiveness in the ability of these disinfectants to perform up to the expectation of the consumer. The way most disinfectants and antibiotics work is by interfering with the metabolic processes of the cells themselves. If you look under an electron microscope at bacterial cells that have been exposed to an antibiotic, the shape of the cell is collapsed, rendering the cell ineffective.

Extensions

1. Have the students collect and grow bacteria from several different sources. In addition to their mouths, have the students collect samples from the following areas:

floors	counters	skin
windows	hair	clothing
toilet seats	sinks	soil
concrete	desktops	sterile swabs

This is obviously not a complete list, but it should get them started. The students can then find out which areas are contaminated the most.

2. Microbiologists (people that study small organisms) that are in "the field" use a substance called agar to grow microbes. Agar is really nothing more that a specially prepared jello (made from seaweed) that bugs like to grow on. The agar can be served up plain or can come with special additions to encourage or discourage specific kinds of bacteria to grow on it. Check with the local university or a high school teacher that is on the ball and see if they can ante up a little assistance.

3. Another great place to go for help or just a good lesson in sterilization is the local hospital or clinic. If you can coerce a surgical nurse or pathologist to visit the classroom with techniques or samples, the students would get a kick out of seeing the real thing in action.

4. In the event that you are feeling ambitious, nab some agar from a scientific supply store and a bag full of disposable, sterile petri dishes. Cook the agar up and add it to the petri dishes by just lifting the edge of the top dish and pouring the agar in quickly. When you cook it up it will be a light amber color and will pour just like the Jello that you cook up for those exciting holiday desserts. I'm still trying to perfect Jello flambe myself. It keeps melting. The agar will set in 20 to 30 minutes, depending on the temperature of the room and the brand that you use. Have the students collect samples from different surfaces using a swab and inoculate the petri dishes.

Once the agar has set, the students can use swabs to collect samples and smear them on the agar surface. After the dishes are inoculated, put them in a warm, dark place to incubate. In all reality, this is just enough information to get you into a lot of trouble. If you want to do this, either call us or get a hold of a good book that explains in detail what you should do.

EXPERIMENT 10

BACTERIA FARM

*Name*______________________________ *Date*_______________

Bacteria should begin to grow on your potato slices. Record the size, shape, color and texture of each colony. Be sure to label your samples.

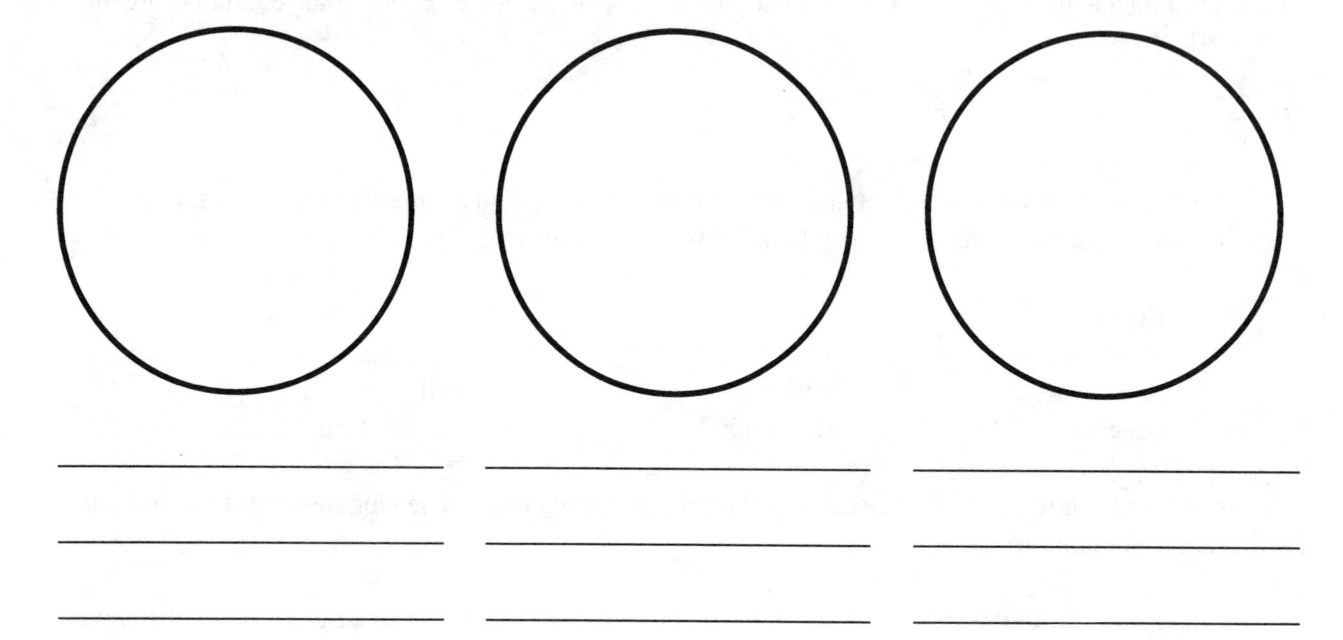

HIDDEN PIGMENTS

Content Area

Plants • Chromatography
Plants • Capillary Action

Learning Objectives

1. The students will observe the separation of pigments from ink and plant samples.

2. They will also infer a correlation between the colors of the pigments found in the separation, and the way the plant uses different colors of sunlight for energy.

Materials You Will Need

Acetone
Mortar and pestle (or spoon and small bowl)
Black felt pen
Different plant specimens
Chromatography paper (or heavy, white drawing paper)
Masking Tape
Pencil or pen
Container or small jar
Water

Procedure

1. Place several leaves in the mortar (bowl), sprinkle in a little acetone (just enough to get it wet with some left over), and grind the leaves to a pulp with the pestle (spoon). A green paste should start to form in the bottom of your bowl in amongst the leaf remnants.

2. Dab a little bit of the leaf paste on the chromatography paper. If you don't have chromatography paper, cut heavy, white drawing paper into strips one inch wide and 10 inches long. The dab of paste should be one inch from the bottom of the strip.

3. Add one half inch of acetone to the bottom of the beaker. Suspend the chromatography strip in the container; the very bottom of the strip should be in the liquid with the dab of paste above the liquid layer. Tape the strip to a pencil that is across the top of the container. This will secure the strip.

Hidden pigments

4. Leave the strip in the container for one to four hours and check the colors of the pigments that rise up the strip periodically.

5. Set up a second experiment using a black spot from a felt tipped marking pen instead of a dab of leaf paste. It's fine if the two strips are side by side in the same container or, if you would like, use water for the felt pen sample.

Whyzat?

The paper absorbs the acetone or water, and it migrates up the strip via a process called **capillary action**. As the acetone comes in contact with the glob of pigment, it literally picks it up and spreads it out along the piece of paper as it goes. The different substances in the pigment are deposited at varying points along the paper, dependent upon how strongly each substance is attracted to the paper. Those with a high degree of attraction are deposited first while those with the least are carried along until they are deposited higher up on the paper.

The different colors give you an idea of the portion of the light spectrum that the plant uses to produce its food. Not all plants use the same wavelengths from the white light for energy and food production, and this is evidenced here. By using different portions of the spectrum, scientists speculate that some plants avoid direct competition with one another.

Extensions

1. The most obvious is to experiment with different kinds of leaves and flowers. All are composites of many different pigments that can be separated out in the chromatography process.

2. You may also want to experiment with the different markers out on the market to see if they are made with the same ink combinations or different recipes. It is also interesting to see what colors comprise other colors of ink, including the fluorescent pens and markers.

3. Changing the solvent that is used is another possibility. The use of this variable requires some degree of common sense. Gasoline, although an excellent solvent, is not a great classroom chemical. Try things like burner fluid (methyl alcohol), turpentine, and mineral spirits. The suggestion here does not mean that the experiment will work, they are just ideas. I would also strongly urge that you use these things in a well ventilated area or outside the classroom if at all possible.

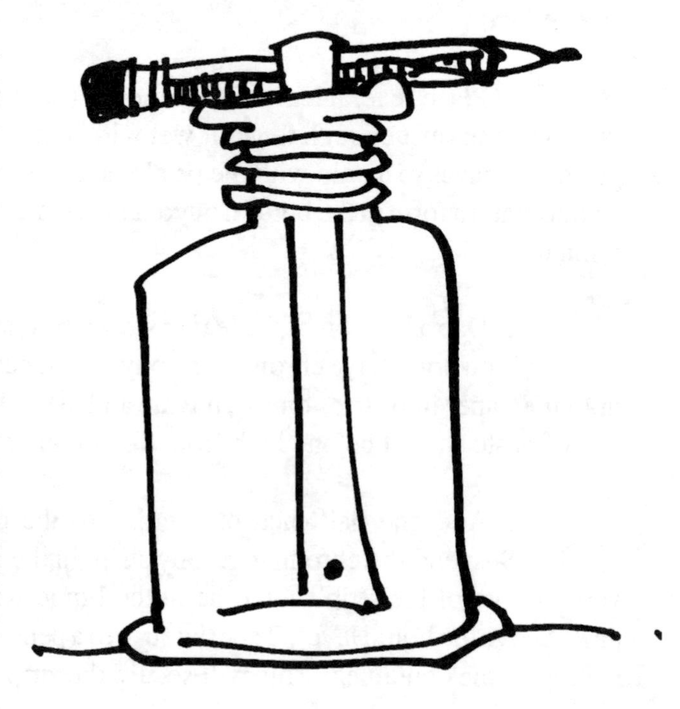

HIDDEN PIGMENTS

*Name*______________________________ *Date*______________

Using colored pencils, draw the pigment scheme that is produced in each of your samples. Label each drawing so you know where the pigments come from.

CHARCOAL GARDEN

Content Area

Geology • Crystal Formations
Oceanography • Coral Reefs

Learning Objective

The students will mix ammonia, laundry bluing, salt, food coloring, water and charcoal to produce a colorful crystal garden through an evaporative process.

Materials You Will Need

Ammonia, household strength
Laundry bluing (hit the local store for this)
Salt
Water
Food coloring
1 Pie tin per lab team
4 Charcoal briquettes per lab team

Procedure

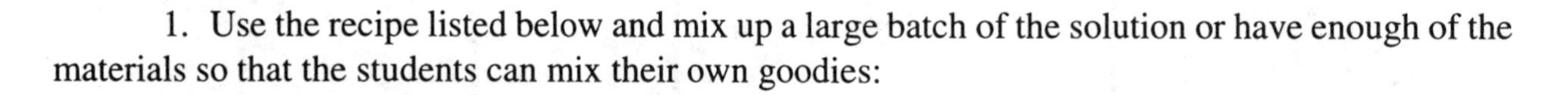

1. Use the recipe listed below and mix up a large batch of the solution or have enough of the materials so that the students can mix their own goodies:

10 parts water
5 parts salt
5 parts laundry bluing
1 part ammonia

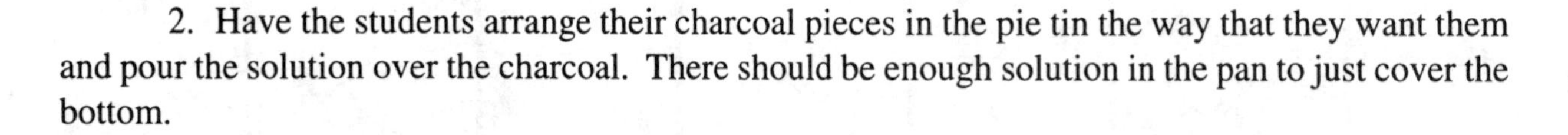

2. Have the students arrange their charcoal pieces in the pie tin the way that they want them and pour the solution over the charcoal. There should be enough solution in the pan to just cover the bottom.

3. The students can add food coloring at this point if they would like, or they can leave the charcoal alone and predominantly white crystals with a blue tint will appear.

4. The crystals will begin to form immediately on the charcoal and also in the pan. As the solution evaporates, add more to the pan. If the students apply the solution directly to the charcoal, the crystals (which are very fragile) will be crushed. In fact, make the students aware of the fact that even blowing very hard on the crystals will knock them over.

CHARCOAL GARDEN

Whyzat?

Charcoal is a very porous material and absorbs the fluid that is in the bottom of the pie tin, as well as some that is poured over the top. Once the solution gets to the top of the charcoal it evaporates, leaving a residue that we see as our crystal garden. The garden will continue to grow until one of two things happen. Either the pie tin will run out of solution, or the crystal will grow too tall to support its own weight and collapse. The different colors are produced by the pigment in the food coloring, which is picked up and carried along with the rest of the solution.

Stalactites and stalagmites in caves are formed in a somewhat different manner. An opening in the Earth's crust is formed under the ground by the movement of large sections of the Earth, called tectonic plates. Incidentally, it can't just be any old opening, it has to be in or near a deposit of limestone. In the process of making the opening underground, the limestone above the opening is fractured by the movement of the Earth. When it rains , the fracture allows water to seep into the rock formation and some of the limestone (calcium carbonate) dissolves. We now have a solution of calcium carbonate (formerly limestone) dripping into an opening that was caused by huge masses of rock smooshing against one another. As the limestone drips into the opening, part of it resolidifies at the point of the opening and some of it drips on to the cave floor and solidifies there, forming stalactites and stalagmites.

Coral reefs are also produced in a similar fashion. Instead of a porous material depositing mineral matter, a coral reef is hundreds of years of coral animals living, growing, and depositing new shells on top of each other. As the shells accumulate (like the mineral deposits accumulate on the charcoal), the reef grows.

Extensions

1. There are all kinds of crystals that you can grow, and if you haven't done the Rock Candy activity in the book (it's the one right after this), I'd say get after that next.

2. A trip to the local museum is always good for reinforcement. There will undoubtedly be a large section that displays minerals and rocks, usually from all over the world.

3. If there is no museum, then there may very well be a rock shop in your area. Ask the owner to come down and bring some of his goodies (not to be sexist, but every rock shop owner I've ever met has been a guy) for the students to look at. Or, if the place is close enough, arrange for a mini field trip.

4. The students may want to experiment with different substances. If you have done the rock candy experiment, it won't take them long to figure out that copper sulfate may work in this experiment, too. Try naphthalene (moth balls) and alum, and, if you don't mind spending a little more money, silver nitrate makes a good batch of crystals. Any good book on crystals will give you several more chemicals that precipitate into crystals.

CHARCOAL GARDEN

*Name*____________________________ *Date*______________

Your crystals will grow each day. Record the growth from the side and the top. Note colors if you have added food coloring. Also note texture and size.

ROCK CANDY

Content Area

Geology • Crystals

Learning Objective

The students will grow rock candy and observe the structure of the crystals that are produced.

Materials You Will Need

Hot plate
4 quart sauce pan (or larger)
Water
Sugar
Cotton string
Food coloring (optional)
Baby food jar or container for growing candy
Masking tape
Pencil or pen

Procedure

1. The standard recipe for rock candy is a 2 to 1 volume ratio. Two parts sugar to one part water. When starting with this solution, be sure to leave two thirds (or more) of the pan empty so that when the sugar is dissolved and displaces the water there is enough room.

2. Fill your pan a little less than one third full of water and set it on the hot plate to boil. Once the water is boiling, slowly add the sugar. Stir to speed up the process of dissolving the crystals. You will add twice as much sugar as you have water and will eventually have a pan of syrup.

3. Once the syrup is made, have the students wet the cotton string in water and then roll it in dry sugar. This "seeds" the string and gives the sugar dissolved in the solution something to hang on to. Have them suspend the string in their container, hanging from a pencil. It should just touch the bottom of the container.

4. The syrup can now be added to the container. Fill it close to the top. If the students want colored rock candy, have them add a bit of food coloring to the syrup and gently swirl it around.

5. Set the container in a place where it will not be disturbed and have the students observe the crystals as they form. It is important that they understand that they are not to pull the string out of the solution to look at it; this disrupts the crystal formation that is going on. You will also want to cover the container because ants love this experiment.

ROCK CANDY

6. Depending on the concentration of the syrup, the number of seeds on the string, and the alignment of the moon, the crystals may take anywhere from 2 weeks to 4 weeks to form. Once they have formed and the observations are complete, the students will want to thoroughly disgust you by eating this concoction of pure sugar; it never fails. I should get a kickback from the American Dental Association for including this experiment in the book.

Whyzat?

One of the basic concepts of science is conservation of energy. Another way of stating that principle is that all things are lazy and like to expend the least amount of energy possible. If you are a water molecule in the gas state you are bopping all over the place. This takes a lot of energy and you'd be using a lot less energy if you were a liquid, so you condense into a water droplet. If the temperature drops low enough, you decide that being in the liquid state takes too much energy, so you turn to a solid and are now a piece of ice or a snowflake. You are conserving energy.

Sugar is usually found as a solid at room temperature, so when it is in the liquid form it's not too happy about all the extra energy it takes to stay there. It's looking for ways to recrystallize. When the students pour the solution into the jar, the sugar molecules use the string as a substrate (base) for recrystallization. I tell the students it's like throwing a life preserver to a man in the water. He goes right over and hangs on. The sugar does the same thing to the string. As more and more sugar molecules hang on, the crystal starts to form.

Extensions

1. There may be a rock shop in your area. Invite the owner to come over and share his or her wares with the class. This is fun for them and good for business, too.

2. If you have done the geode experiment, it won't take them long to figure out that cupric sulfate may work in this experiment, too. Naphthalene, silver nitrate, and alum also make good crystals. Other ideas, you say? Ok, how about potassium permanganate, cupric nitrate, sodium silicate, lead nitrate, and sodium thiosulfate. If you want more ideas check with your local chemist.

3. Finally, a museum will have a large collection of minerals for the students to peek at and oooh and aaah over. There is usually a gift shop that will sell minerals as well, so you can pick up a couple of different samples for the class.

4. You'll find the suggestions for this lab activity to be quite similar to that of Charcoal Gardens and Eggshell Geodes. They all reinforce each other.

ROCK CANDY

*Name*______________________________ *Date*______________

Record the growth of the crystals each day. Be careful not to remove the crystals from the solution; this interrupts the growth process.

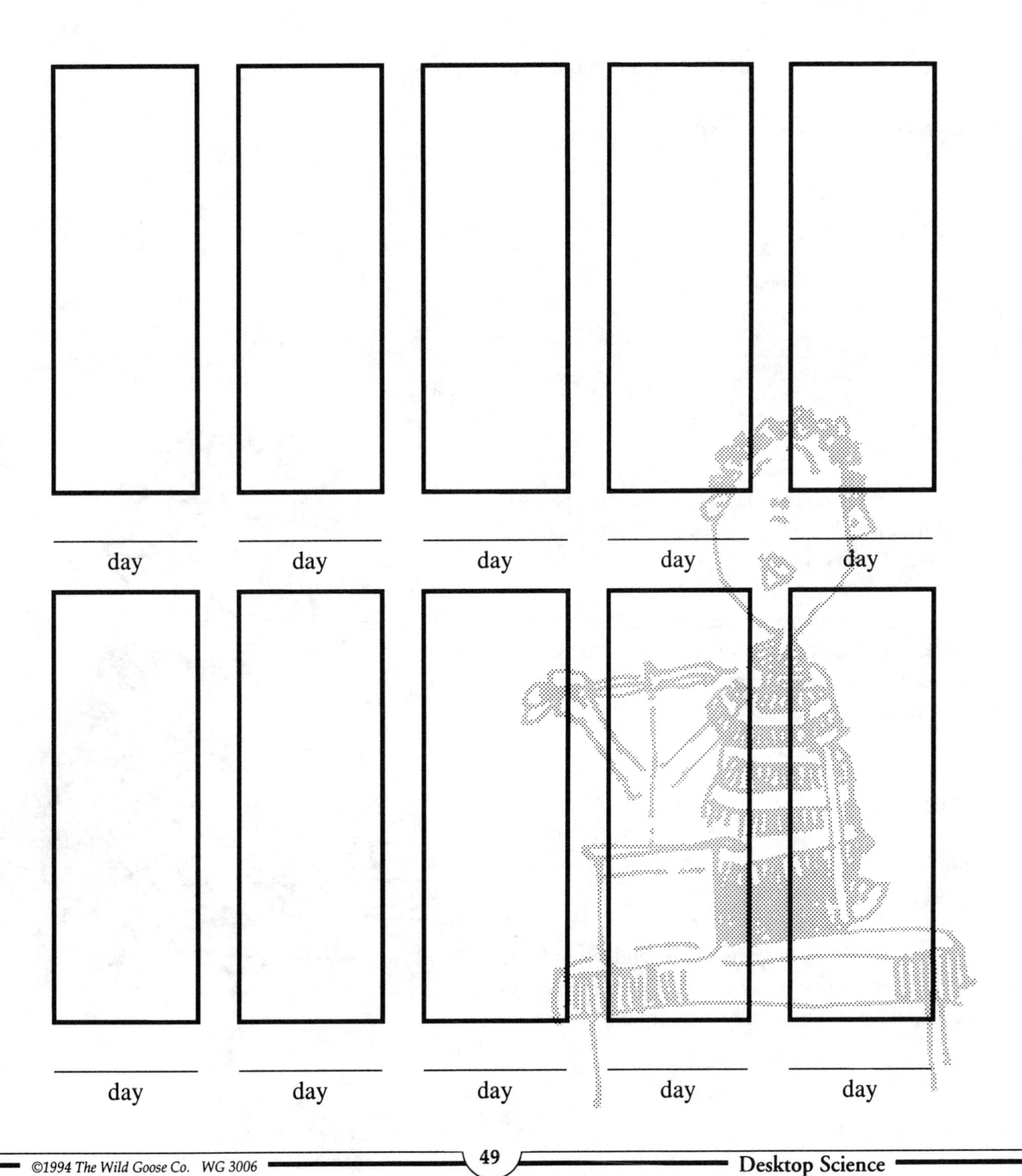

DISAPPEARING ACT

Content Area

Weather • Evaporation
Geology • Formation of Evaporative Rocks

Learning Objective

The students will observe and record the evaporation of salt water.

Materials You Will Need

Water
Container
Salt
2 Pie tins per lab team
Food coloring (optional)
Masking tape
Pencil or pen

Procedure

1. Using the masking tape, label pan #1, *salt water* and pan #2, *tap water*. Put the label in the pan, not on the bottom.

2. Add enough water to pan #2 to just cover the bottom and set it aside. Record the amount of water that you use, and use the same amount to prepare pan #1.

3. Add a tablespoon of salt to the water that you are going to pour into pan #1. Stir it until it is completely dissolved and continue to add salt until no more will dissolve. After it is saturated, pour it into pan #1.

4. Set both pans aside and record the observations every day until all of the liquid in both pans has completely evaporated.

DISAPPEARING ACT

Whyzat?

The water evaporates from the pan. Because the chemicals are in **solution** (mixed in with, but not attached to, the water molecules) only the water is evaporated, leaving the mineral residue. I think this is the shortest whyzat in the book, but that's all there is to the experiment. Actually, this is the same process that produced the charcoal crystal garden.

Extensions

1. Have the students vary the concentration of the salt water and see if that will affect the rate of evaporation. Be sure that they use the same amount of water each time and only vary the amount of salt used.

2. They may also want to vary the solvent used. Have them try vinegar, ammonia, etc. When they do this experiment, make sure that they vary only one thing, the solvent. This means that they should use the same amount of salt and vary the liquid.

3. A comparison of temperature will provide meat for a discussion. Place one container in the sun, one in the shade, and one in the refrigerator. Or place one in the sun inside the room and one in the sun outside the room. Do the same thing with the shaded parts of the room and outside.

4. There is a whole group of sedimentary rocks that are formed this way. They include, among others, gypsum, halite (rock salt), and potash. It is possible that you may have a processing plant for one of these three rocks near your area, especially if you live in the west.

A field trip is in order or a visit from a mining engineer who works at the plant and would be willing to come and chat.

DISAPPEARING ACT

*Name*____________________________ *Date*________________

Observe the evaporation of the solutions from the two dishes. Draw what you see.

Salt Water **Tap Water**

Time ________

Time ________

Time ________

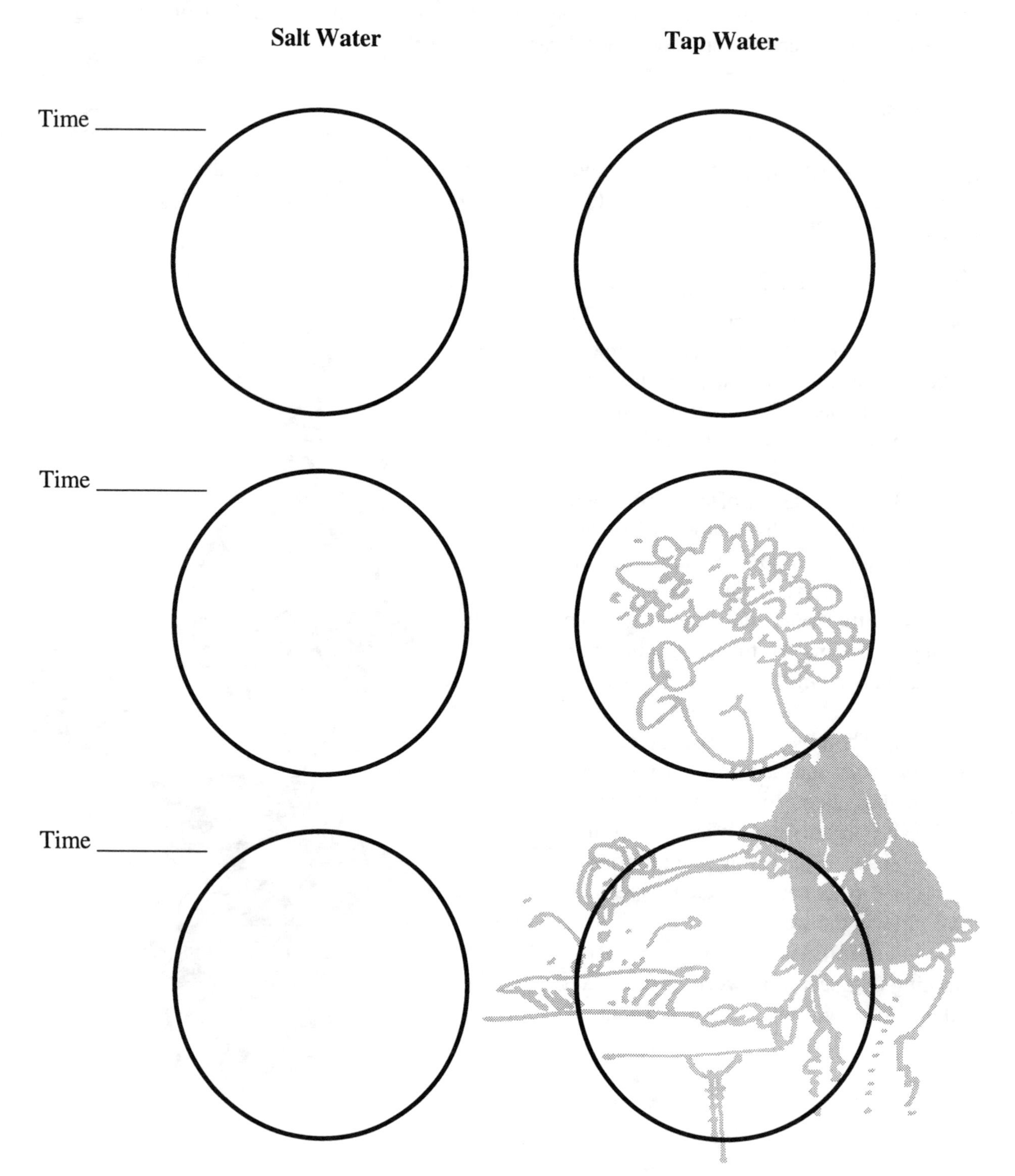

EGGSHELL GEODES

Content Area

Geology • Geodes
Weather • Evaporation

Learning Objective

The students will prepare a solution of copper sulfate and record observations as it evaporates and produces crystal patterns in the bottom of an eggshell.

Materials You Will Need

Egg carton per lab team
Eggshell halves (empty)
Water
Copper sulfate crystals, 1 pound
Container, metal pan
Spoon
Hot plate

Procedure

1. Add a half quart of water to your container and place it on the hot plate.

2. Once the water is boiling, begin to add copper sulfate crystals slowly. Stir the solution constantly, adding additional crystals as the others dissolve. Add the copper sulfate until the crystals won't dissolve anymore. You have created a super-saturated solution.

3. Have the students select clean eggshell halves and place them in egg cartons for support. Pour the copper sulfate solution right to the brim of the eggshell.

4. Carefully set them aside and observe as the solution evaporates. When the solution has completely evaporated, you will have something that resembles a geode.

EGGSHELL GEODES

Whyzat?

As the water in the copper sulfate solution evaporates, the chemical is left behind. Many solids form characteristic patterns when they are not in solution. The best way to describe this would be to compare the copper sulfate to a set of square building blocks. If you toss the building blocks into the river, they float around and bop all over the place in no particular order. When the river dries up the blocks come to rest. Because of the shape of the blocks they are likely to form a square (I realize we are really stretching this analogy here; bear with me). Copper sulfate is likely to form the crystal shape that you see in the geode because that is the way those chemicals tend to line up.

A geode is a mineral deposit that is formed when a gas bubble is trapped in a lava flow. Over time the bubble fills with mineral water and evaporates, fills and evaporates. Each time this happens the crystals begin to form. Later in the geologic time period the lava flow starts to erode. As it is eroded, the geodes, which are now harder than the surrounding lava, erode out as well and are left for us to find.

Extensions

1. If a trip to a local museum is in the offing, there will undoubtedly be a large section that displays minerals and rocks. The students will find geodes in the collection.

2. If there is no museum, then there may very well be a rock shop in your area. I can remember, as a child growing up in Portland, Oregon, wandering down to the rock and lapidary shop on Sandy Boulevard and spending hours peeking at the rocks and minerals the old guy had in the bins. If you are lucky enough to get a social creature that owns one of these shops in your area, invite him or her to drag his or her goodies down to your class for show and tell.

3. The students may want to experiment with different substances. If you have done the rock candy experiment, it won't take them long to figure out that sugar may work in this experiment, too. I can't recommend any other chemicals at this time, but if you get hold of a good book on crystals or even that famed character, Mr. Wizard, I'm sure either resource will have additional ideas.

4. If you are into the exotic, have the kids bring in or start a collection of geodes that are produced and found around the country. They are common throughout the west in areas of volcanic activity. Utah has septarium nodules as well as their Dugway Geodes. The state rock of Oregon is the Thunderegg. Idaho and Nevada have several rich collecting sites and the imported geodes from Brazil and the South Pacific are spectacular as well as expensive.

EGGSHELL GEODES

*Name*________________________________ *Date*______________

In the spaces provided below, draw 1) the empty egg carton, 2) the egg cartons full of copper sulfate solution, and 3) the finished experiment. Add your observations of the finished crystal.

1.

2.

3.

Observations____

____

____

INSTANT ROCK

Content Area

Geology • Sedimentary Rocks

Learning Objective

The students will use glue, water, and sand to make a sedimentary rock. They will then experiment with other kinds of adhesives and compare the results.

Materials You Will Need

Sand
Water
White glue
Clean, empty milk carton/lab team (1 pint size)
Various adhesives
Masking tape
Pencil or pen

Procedure

1. Prepare a solution that consists of one third water and two thirds white glue. Stir the concoction well.

2. Clean an empty milk carton and fill it with sand. Leave about a half an inch at the top.

3. Add the mixture of glue and water until it comes just to the top of the sand. Let the entire mixture set for one full day; two is better if you live in a humid climate.

4. Try additional substances for the same effect. Household items are always amusing; flour, cornstarch, sugar, and baking soda come to mind. Mix up at least one extra rock, or several if you have the sand and the milk cartons.

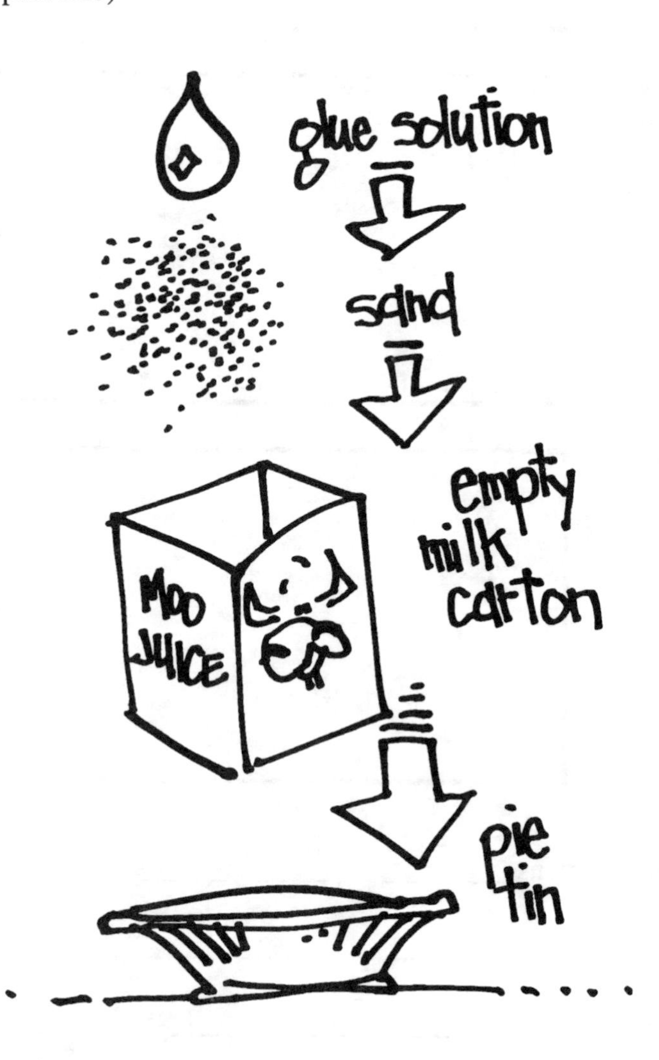

INSTANT ROCK

5. After the rocks have set up, peel the carton away from the sand and test the different rocks for strength. You may want to save a couple for open house, back to school night or Charlie Brown, if it's close to Halloween.

Whyzat?

Some kinds of sedimentary rocks are formed over long periods of time. Streams and wind deposit sand, silt, and dead plant and animal matter (these turn to fossils if they are buried quickly enough) in layers. As the years pass, the layers get deeper and deeper. As you can imagine, the weight of all of this muck smooshes* the sand at the bottom, cementing it together to form a **sedimentary rock**, or rock made of smooshed silt and sand (sediment). The different colors are caused by the presence of different kinds of minerals and ores. Red is usually indicative of iron, blue of copper, black of manganese and so on.

The different adhesives have no direct scientific application that I am aware of, but it is a lot of fun to see what can happen to the stuff we put in our bodies.

* Lest there be any confusion or dissent, "smooshes" is a perfectly acceptable and descriptive scientific term in elementary science. Use it often if you like, and feel free to make up others as you go. The objective is to teach concepts and ideas. They can be clinical when they get to high school.

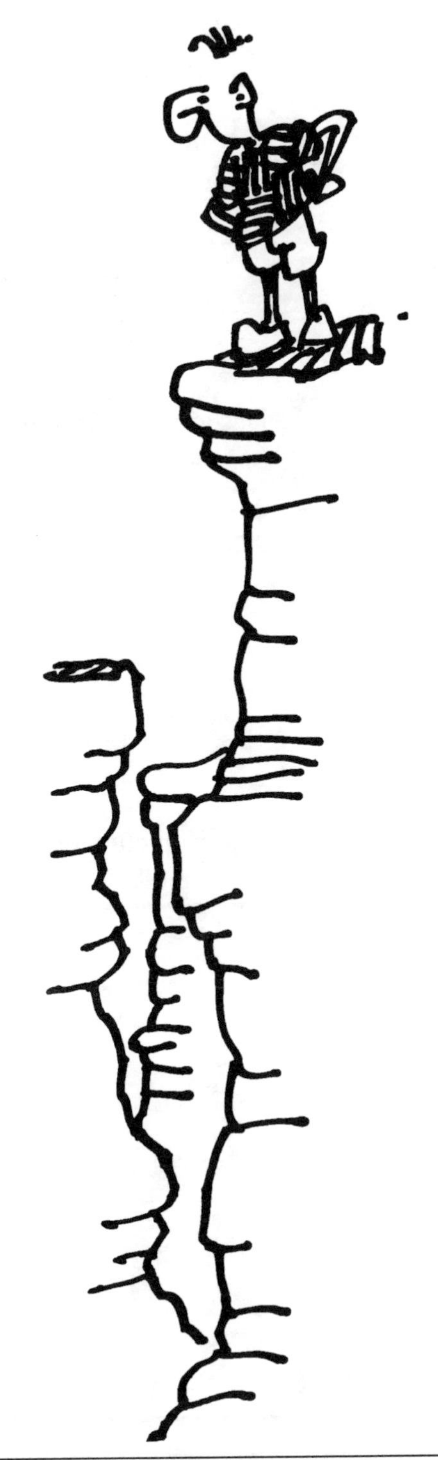

Extensions

1. Try as many different kinds of adhesives as you can find, would be my suggestion. Once you have done that, try different materials; "rocks" can be made out of dang near anything.

2. Sedimentary rocks abound in formations throughout the United States. A trip to the local museum of natural history might be in order. Or, if there is a rock shop in the area, have the local owner come up and show off his goodies; both are good extensions.

3. For the ultimate in sedimentary rock formations, plan a trip to the Grand Circle in Utah, Arizona, New Mexico, and Colorado. It will bug out your mind.

INSTANT ROCK

*Name*______________________________ *Date*______________

Fill in the spaces below.

1. Describe the sand before you began.____

____

____

____

2. Describe the finished "sedimentary rock". _____________________________

____

____

____

____

3. Draw a picture of the rock in the space below.

HOMEMADE THERMOMETER

Content Area

Weather • Thermometers
Physics • Expansion and Contraction of Fluids

Learning Objective

The students will make and calibrate a homemade thermometer and use it to record the temperature of various environs.

Materials You Will Need

- 2 Containers (drinking glass size)
- 1 Pyrex test tube per lab team
- 1 One hole stopper per lab team
- 1 12" section of glass tubing
- 1 Test tube holder per lab team
- 1 Grease pencil per lab team
- 1 2" x 8" white card per lab team
- Water (Distilled)
- Food coloring
- Pencil or pen
- Hot plate
- Sauce pan
- Bowl
- Ice
- Salt
- Rubber bands or tape

glass tubing
1 hole stopper
test tube
water w/food coloring

Procedure

1. Plug the hot plate in, place the sauce pan full of distilled water on it and fire it up. Get the water boiling. Place the ice cubes in the bowl of distilled water and make sure it is as cold as it can possibly get.

2. Fill the test tube full of water and add a pinch of salt and a couple of drops of food coloring for effect. Insert the glass tubing in the stopper and insert the stopper in the top of the test tube. Use the diagram at the upper right as a guide. The colored water should stick up the tubing one third of the way from the top of the stopper. If it is higher or lower, adjust it and then mark this spot with the grease pencil as the beginning level.

HOMEMADE THERMOMETER

3. Attach the test tube to the white card, using the rubber bands or tape. Mark the spot on the card where the top of the glass tube is found. This is your reference mark when you replace the card. Remove the card.

4. Put the test tube in the test tube clamp and place it in the cold, distilled water. Let the test tube equilibrate (adjust) for at least one minute (longer is better; in this case, up to 3 minutes), then pull it out of the water and quickly mark the water level on the card. Be sure to have the students line up the top of the tube with the mark on the top of the card before they put the second mark on the card.

5. Once that mark has been made, remove the card again and place the tube in the boiling water. Let the tube equilibrate again. Replace the card and mark the spot on the card. Remove the card again.

6. The two marks represent 0 and 100 degrees on the Celsius scale. The task at hand is to measure the distance between the marks in millimeters, divide by 10, and make a mark to represent 10 degrees Celsius. For example, if the distance between the top and the bottom of your marks is 120 mm, then you would place a mark representing 10 degrees every 12 mm. There is a worksheet to help you figure this out.

7. When the students are done, they can approximately measure the temperature of various substances and environments with their instruments.

Whyzat?

Everything in the world is made up of those now famous, tiny particles called **molecules**. At room temperature they are bumping and bouncing and rolling past one another constantly. The warmer they get, the more energy they have and the more space they need to bump around. The colder they get, the less energy they have and the slower they move, so they need less space.

We put these principles to work to make the thermometer. As the water that is trapped in the test tube is heated, the particles have more energy and want to take up more space. They expand up the tube. As the water is cooled, the molecules have less energy and require less space. They contract and withdraw down the tube. By knowing the boiling point and freezing point of water, and measuring the distance between the two, we can predict the temperature of different objects fairly accurately once a scale is constructed.

There are two substances that are commonly used in commercial thermometers. They are mercury, the silver liquid in the more expensive thermometers, and alcohol, the red liquid in the less expensive thermometers.

HOMEMADE THERMOMETER

Extensions

1. Have the students make thermometers using different fluids. Every substance that is on the planet has what is called a **coefficient of thermal expansion**. This coefficient is a number that describes how much that material will expand as it is heated. To use nonscientific terms, if a substance has a big number it expands a lot when it is heated, and if it has a small number it does not. Different fluids will have different coefficients or rates of expansion, but all can be used to make a thermometer.

2. If you want to experiment with temperatures colder or warmer than the range provided between the freezing and boiling points of water, you will need to make sure the fluid you use in your thermometer will accommodate you (i.e. you don't want your thermometer to freeze or boil). For example, salt was added to your thermometer fluid to measure colder temperatures because salt water has a lower freezing point. Remember though, it is important to realize that you must calibrate your salt water thermometer with the same method, and the same fluid (i.e. pure, distilled water), shown in the Procedure section; otherwise, you won't know the actual temperatures (zero and 100) of the two calibration points.

3. What can you measure that is colder than zero? How about salt water similar to that which is now in your salt water thermometer?

4. To experiment with warmer temperatures, you might use ordinary cooking oil as your thermometer fluid since its boiling point is well above one hundred degrees Celsius. Once again remember, place this oil thermometer in freezing and boiling water to calibrate it. If you end up using your oil thermometer to measure the temperature of a sauce pan of oil, make sure you heat it slowly and use a great deal of CAUTION in the process.

5. Introduce the idea of three temperature scales, Fahrenheit, Celsius, and Kelvin. Each one has a different application and use. A little research and you will be up to speed on all three and a veritable thermometer expert. Just what you always wanted to be when you grew up.

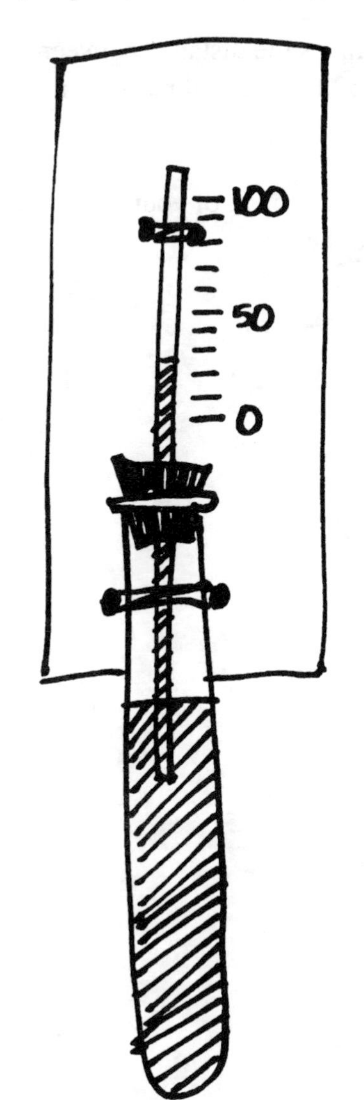

HOMEMADE THERMOMETER

*Name*______________________________ *Date*______________

1. Use the following information to build your thermometer scale:

a. Distance between 0 and 100 marks. Measure in millimeters. __________ mm.

b. Divide this number by 10

$$\frac{\_\_\_\_\_\_\_\_\_\_\text{ mm}}{10} = \_\_\_\_\_\_\_\_\_\_$$

c. This is the distance between every 10 degrees. Measure and mark them on your card.

2. Make a replica of your temperature card in the space to the right.

SUNDIAL

Content Area

Astronomy •Sundial

Learning Objective

The students will make a sundial and record the movement of a shadow around it during the course of a day and over the period of a month.

Materials You Will Need

Cardboard (8"x 8")
Straight pin

Procedure

1. Draw a pattern similar to the one pictured below on one side of the cardboard.

2. Insert the pin in the center of the cardboard from the backside.

3. Check the time of day on the classroom clock and place your sundial outside in a sunny area, adjusting the sundial so that the shadow falls on the correct time and taping it down so that it is difficult to move. Note the location of the shadow when you took the reading. Make a small "x" on the cardboard at the tip of the shadow, and write the date next to the mark.

4. Every hour for the rest of the day, make an "x" at the tip of the shadow. Compare the sundial time with the actual time to see how accurate your new "clock" is.

5. Repeat the previous two steps once a week for at least the next four weeks; the more the merrier, twelve weeks would be great. Observe how the length of the shadow changes and decide how a sundial could be used to tell the date in addition to the time.

EXPERIMENT 18

SUNDIAL

Whyzat?

The sun's position in the sky changes during the course of the year due to two reasons: 1) the earth is tilted on its axis, and 2) the earth is in a constant orbit around the sun. The sketches on the next page help to illustrate what is happening.

The first sketch shows the earth at four positions in its orbit around the sun, all the while tilted on its axis. Notice how the tilt of the earth never actually changes—it just seems to change due to the earth's movement around the sun. Pretend like you are in a spaceship trailing along behind the earth, so that the sun is always on your left, and you will observe the "apparent" changes of the tilt relative to the sun. Spaceship positions are shown by "x" and are labeled as A, B, C and D.

The next four sketches show "views" from the spaceship at each of the four positions A, B, C and D. What you are viewing is a giant needle (roughly five hundred miles tall) sticking straight out of the earth at the 45th parallel and its shadow. Each view is assumed to be made at twelve noon.

Sketch A shows what you would see on the winter solstice, the shortest day of the year, when the sun is at its lowest position in the sky. Notice the long shadow cast by the giant needle.

Sketch C shows the other extreme, the summer solstice, where the sun is at its highest point in the sky, and the daylight hours are the longest. This position of the earth in its orbit provides for the shortest shadow from the meganeedle.

Sketches B and D show in-between positions: the vernal equinox (first day of spring), and the autumnal equinox (first day of autumn).

The sundial gives the exact same results. The only difference is a one inch needle versus the five hundred mile needle.

Extensions

Challenge the students to make a more accurate sundial or to make a sundial that can tell the time by the hour. If you have a perfectionist in your class or a potential engineer, have them record the movement and position of the sundial for the entire year. As the seasons change, the length and position of the sundial shadow will also change.

Have one of your more ambitious students figure out how to construct a gnomon and build a sundial using the information that they find in the library.

SUNDIAL

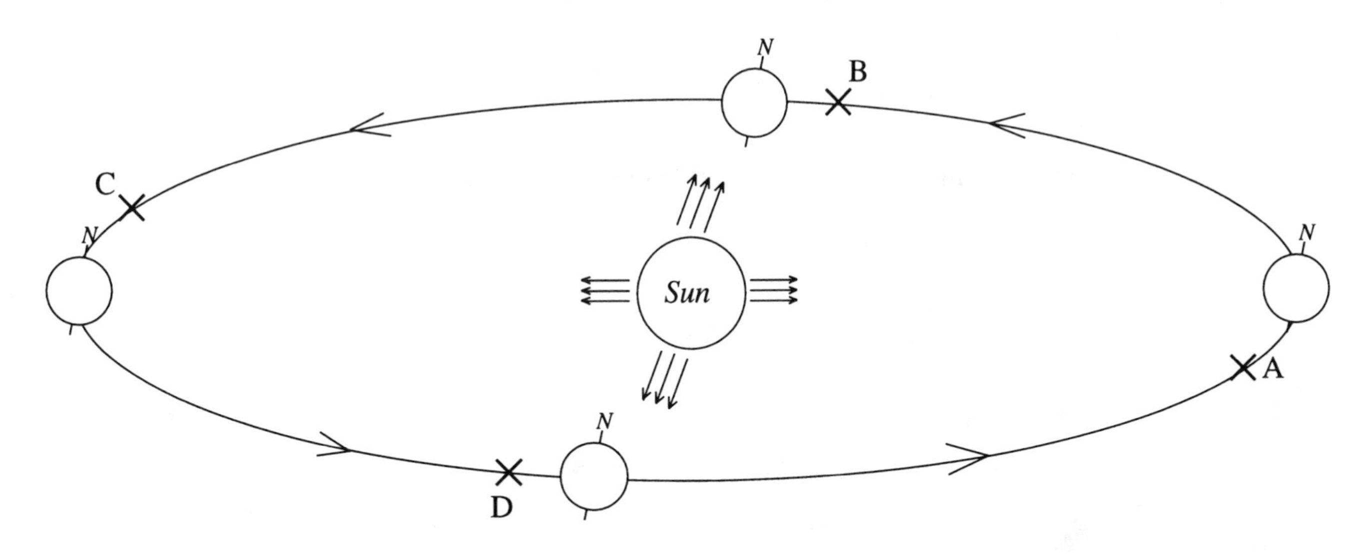
N
B
C
N
Sun
N
A
N
D

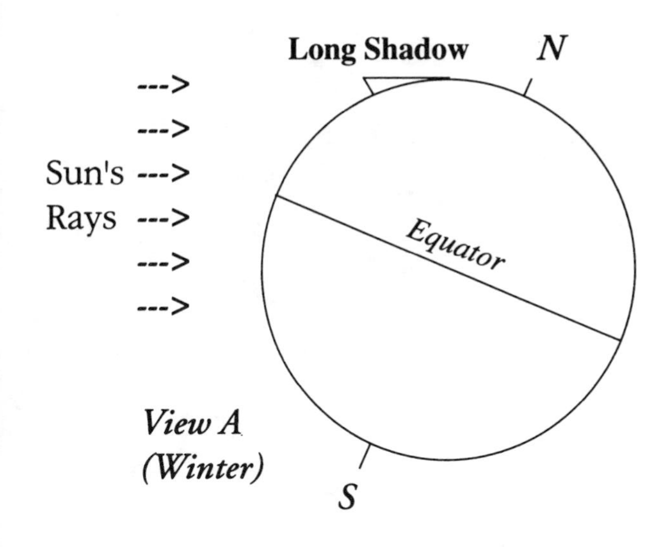
Long Shadow
N
--->
--->
Sun's --->
Rays --->
--->
--->
Equator
View A
(Winter)
S

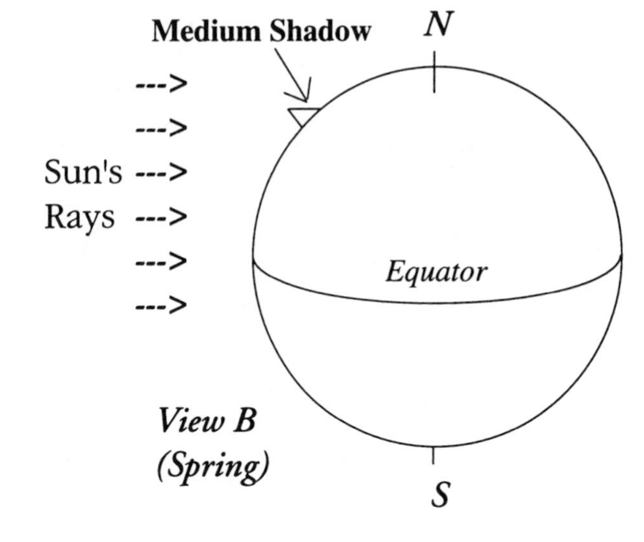
Medium Shadow
N
--->
--->
Sun's --->
Rays --->
--->
--->
Equator
View B
(Spring)
S

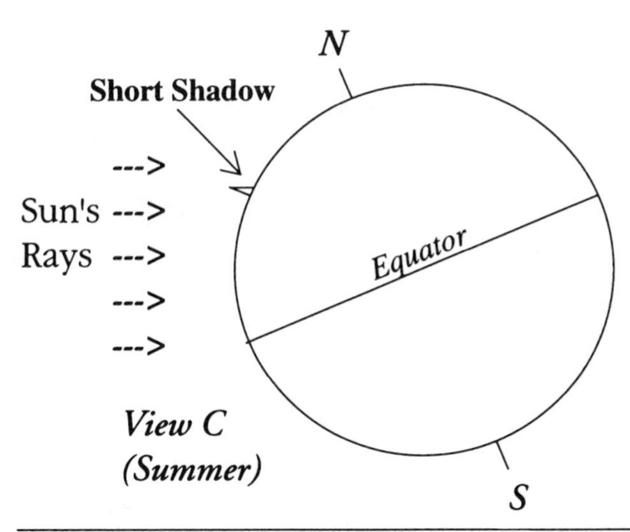
N
Short Shadow
--->
Sun's --->
Rays --->
--->
--->
Equator
View C
(Summer)
S

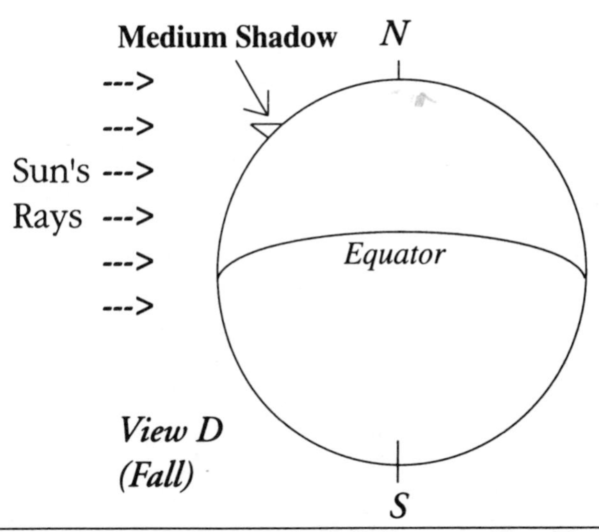
Medium Shadow
N
--->
--->
Sun's --->
Rays --->
--->
--->
Equator
View D
(Fall)
S

SUNDIAL

*Name*______________________________ *Date*______________

Detail a design for a more accurate sundial in the space below. Include a drawing.

EXPLODING MEDICINE

Content Area

Health • Time Release Medicine

Learning Objective

The students will chronicle the release of medicine from Contac capsules.

Materials You Will Need

1 Contac® capsule per lab team
1 Container per lab team
Water

Procedure

1. Give each student a Contac® capsule and ask them to gently separate the container and dump the medicine on the table. Ask them to count the number of medicine pellets.

2. Once they have counted the number of medicine pellets, have them put them in the water. They should all sink. The rest is clinical.

3. Every half hour for the next 6 hours, have them count and record the number of medicine shells that have exploded and are now floating on the surface of the water. If you have longer than 6 hours, have the students go for it.

4. When they are all done, have them determine if the medicine is released evenly or in irregular doses.

Whyzat?

The medicine is encapsulated in tiny balls that are coated with time release coverings. Each covering is designed to absorb water. When the covering has absorbed enough water, it gets too big to hold itself in and it bursts, releasing the medicine to be absorbed by the individual's body.

The idea is to have the coverings release the medicine at different rates so that the patient doesn't have to constantly take medication, especially at night. As we used to say in the sixties, "Better living through chemistry." A slightly different context these days, perhaps.

EXPLODING MEDICINE

Extensions

1. Temperature is the most obvious variable here. Have the students experiment with different temperatures and see if the rate of release is affected. Have them experiment with water that is near freezing and water that is near, but not quite, boiling.

2. Also, the pH of the human body is very specific. Make an acidic solution by adding vinegar to the water, and a basic solution by stirring in baking soda. See if this influences the rate of release.

EXPLODING MEDICINE

*Name*____________________________ *Date*_______________

In the data table below, record the number and color of exploded casings that are floating on the surface of the water.

TIME	COLOR	# OF SHELLS
1 hour	yellow	
	white	
	red	
2 hours	yellow	
	white	
	red	
3 hours	yellow	
	white	
	red	
6 hours	yellow	
	white	
	red	

Is there a correlation between number of capsules and colors of casings? __________

MEALWORMS

Content Area

Insects • Mealworms
Insects • Metamorphosis

Learning Objective

The students will watch and describe the changes a mealworm makes as it grows through its larval stages and during the metamorphosis from a pupa to an adult beetle.

Materials You Will Need

1 Baby food jar per student
1 Mealworm per student
Apple or potato
Bran
Masking tape
Pencil or pen
Water

Procedure

1. Give each student a baby food jar and ask them to put their name on it.

2. Set up a buffet-style, mealworm adoption center. Place the bran in a bowl or tin plate first, then the sliced up apple or potato, and finally the mealworms themselves.

3. Have the students come by and take the items in the order they are listed in step #2 and place them in their jars. When they are all done, have them take their new critter back to their desk, name it, and make the initial observations. It is worth noting that some of your students will exhibit an unusual fear toward this small, harmless creature. (This is due primarily to a parent who doesn't know much about insects themselves.) To alleviate this fear, I simply sit the child down, show them a mealworm and then pop it into my mouth. This rarely does anything constructive, education-wise, but it really is a lot of fun and the student usually gets the idea. I tell them that mealworms are very friendly, are very curious and want to get to know their new owner, are very funny because they like to tickle your hand and love to be treated with TLC. This usually does it. If it doesn't, I don't push it.

4. Have the students continue to make observations for the next 3 weeks as the larvae change to adults.

MEALWORMS

Whyzat?

The largest group of animals is the class insecta, or the insects. It has been estimated that there are more insects than all other animal species combined. This is not too hard to believe since the roll books have 800,000 different species of insects in attendance (275,000 of which are beetles). In fact, one of the great fears that comes with the destruction of the rain forests in Central and South America is the potential loss of hundreds or thousands of kinds of animals including all those cool, new bugs hanging out waiting to make the pages of the encyclopedia or the next Stephen King novel.

Back to the serious stuff. Most insects go through four distinct phases in a process called complete metamorphosis. They all begin as an egg. Once the eggs hatch, they enter the larval stage. Depending on the bug, you may know this stage as maggots, caterpillars, grubs or, in our case for beetles, mealworms, but the catchall term is **larva** (lar • vuh). The larva will molt or shed its skin a couple of times, and during the final molt it appears to take on an outwardly, lifeless form called the **pupa** (pew • puh). While the insect is in the pupal stage, it changes into its adult form. Once an **adult**, it can then reproduce and lay more eggs to start the cycle again.

egg

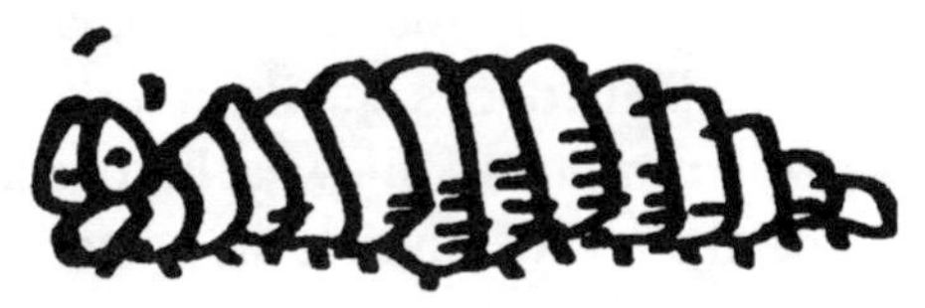

larva

The eggs of the mealworm are quite small and difficult to see without a magnifying lens. When you buy the mealworms from the pet shop or garden store, they will come to you in the larval stage as cream colored, smooth, semi-caterpillar-like creatures without the fuzz. As they grow, they will molt several times and progress through successively larger larval stages called **instars**. The bug is simply getting bigger. During its final instar, however, it becomes a pupa. It is entirely possible that your students will announce to you that the mealworms have died. This is called an instructional opportunity. Explain that the insect is merely changing its outfit, and, being the modest creature that it is, it would prefer to do so in private. Once the metamorphosis is complete during the pupal stage, the adult insect will emerge. In this case our mealworms have become black beetles.

pupa

If you keep the mealworms in a container full of flaked bran or oatmeal with apple slices as a source of water, they will be quite happy, reproduce, and provide you with an endless supply of mealworms. Pretty nice set up, if you ask me. And if you are stumped when it comes to the fund raiser for next year you have a built in revenue generator.

adult

MEALWORMS

Extensions

1. If your classroom lags in the excitement department, there is nothing like a good, old-fashioned beetle race to stir the blood and coagulate the fears of entomophobics. Draw a one foot diameter circle on the linoleum (or blacktop). Draw another circle three feet in diameter around the inner circle. Have the students place their beetles in the middle, then let the action begin. It's actually more fun to watch the students than the beetles.

2. There are all kinds of animals that go through life cycles. When I was teaching in Southern California, the mulberry trees were denuded every year for a pile of silkworms. Butterflies can be ordered in either the larval or chrysalis stage. If you live out a ways in a wet area (Houston comes to mind for some reason), skim the tops of the ponds in the spring and you'll come up with mosquitoes, dragonflies, damselflies, and who knows what else in the larval stage. Also, if you go to streams in the spring and turn over rocks and branches, you'll find all kinds of insect larvae attached to the bottoms of the rocks and sticks in most any stream.

3. If you are into the bigger and more adorable, try tadpoles. Or, if you have the equipment, chickens, ducks, quails, or turkeys. One last resort, which can actually be a lot of fun, is to get a pregnant mom, who is somewhat literate and has some degree of communicative skills, to come to your school and show off during the gestation of her child. If you have a shortage of these, let us know because we have more than our fair share of pregnant moms here in Utah. Must be the water.

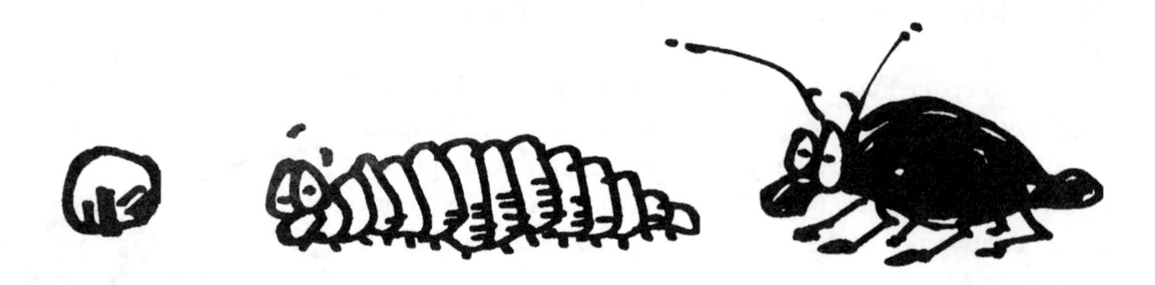

MEALWORMS

*Name*______________________________ *Date*______________

Record the following information.

Mealworm

Number of body parts ______________________

Number of feet ______________________

Length ______________________

Width ______________________

Color ______________________

Movement ______________________

Pupa

Number of body parts ______________________

Number of feet ______________________

Length ______________________

Width ______________________

Color ______________________

Movement ______________________

Beetle

Number of body parts ______________________

Number of feet ______________________

Length ______________________

Width ______________________

Color ______________________

Movement ______________________

THE SCIENTIFIC METHOD

1. Think of an Idea

The first thing that you will need to do is think of an idea for what you will try to explain or do in your experiment, or just something you may want to study. The best way to get started is to adapt an existing experiment in your own unique way.

2. Research Your Topic

Find out what is already known about the topic, and see what you can add to the general body of knowledge.

3. Plan Your Experiment

This section is also called the procedure. You make a game plan of when, where, how, what, and why you are going to do what it is that you are going to do.

4. Experiment

Party time. This is where you get right down to the nitty gritty of doing the experiment, collecting the data, rolling up the sleeves and diving right into the fray.

5. Collect and Record Data

This is all the information that you are seeking—including charts, data tables and records of observations.

6. Come to a Conclusion

Compile the data that you have collected, evaluate the results, come to a conclusion, write a law describing what you observed, and collect your Nobel Prize.

STEP 1

Think of an Idea

STEP 2

Research Your Topic

Step 3

Plan Your Experiment

STEP 4

Experiment

STEP 5

Collect & Record Data

Step 6

Come to a Conclusion